THIN PLACES

◊

THIN PLACES

A Pilgrimage Home

ANN ARMBRECHT

COLUMBIA UNIVERSITY PRESS   NEW YORK

25.00

COLUMBIA UNIVERSITY PRESS
Publishers Since 1893
New York   Chichester, West Sussex

Library of Congress Cataloging-in-Publication Data

Armbrecht, Ann.

Thin places : a pilgrimage home / Ann Armbrecht.

p. cm.

Includes bibliographical references.

ISBN 978-0-231-14652-4 (cloth : alk. paper)

ISBN 978-0-231-51829-1 (e-book)

1. Yamphu (Nepalese people)—Nepal—Hedanga—Social life and customs.

2. Hedanga (Nepal)—Social life and customs.

3. Armbrecht, Ann.

4. Women anthropologists—United States—Biography.

5. Women anthropolgists—Nepal—Hedanga—Biography. I. Title.

DS493.9.Y36A76 2009

305.895′49—dc22

2008023537

Columbia University Press books are printed on permanent
and durable acid-free paper.

Printed in the United States of America

c 10 9 8 7 6 5 4 3 2 1

DESIGN BY MARTIN N. HINZE

IN MEMORY OF JULIE BETH GOLDMAN

DECEMBER 31, 1962–SEPTEMBER 18, 2000

◊

It is only with the heart that one can see rightly;
what is essential is invisible to the eye.

ANTOINE DE SAINT-EXUPÉRY, *The Little Prince*

There are moments when the walls of my mind grow thin;
when nothing is unabsorbed.

VIRGINIA WOOLF, *The Waves*

◊

. . . there is no place
that does not see you. You must change your life.

RAINER MARIA RILKE, "Archaic Torso of Apollo"

CONTENTS

◊

Many years ago, two brothers stood at the edge of an icy-blue lake high in the Himalayas in a pass marking the border between Nepal and Tibet. These brothers, called Minaba and Sepa, were the sons of Yamphuhang, a Kiranti prince who had been chased with his brothers from their father's kingdom in Kathmandu by Mughal invaders from the south. Yamphuhang and his brothers had fled to the eastern reaches of their father's land, where they then headed north, following different river valleys in search of the lands they would eventually claim as their own. Yamphuhang, the second son, chose the middle valley, that of the Arun River. He traveled for days, wrapping his arms and legs in the bark of trees to keep warm, climbing through the mountains until he eventually reached the Tibetan Plateau. He settled, married, and fathered two sons: Minaba and Sepa.

Once Minaba and Sepa were old enough to claim their own land, they decided that the arid Tibetan Plateau was too harsh and the life too hard. And so they had come to this lake in a pass called the Popti La. Standing at its edge, they threw a wooden bowl lined with silver and a walking stick into the lake. They watched as the bowl and stick spun slowly around the center of the lake and disappeared beneath the surface. And then they swore that wherever the bowl and walking stick resurfaced, that place would be their home.

"Singing, singing, looking, looking," they began to climb down the mountains. The two brothers flew like birds, the stories say, soaring over the narrow ravines and steep ridges of those rugged lands. They traveled for days—"walking, walking, watching, watching"—stopping only when they finally saw the reflection of a lake through the forest. Minaba and Sepa climbed a ridge to get a better view. The bowl and stick circled in the center of the lake. The brothers looked around. Huge boulders covered the steep slope. Life did not look any easier in this rocky land than it had on the high Tibetan Plateau. They were being led by the ancestors, but certainly they had some say in the matter. They threw the bowl and stick back into the water and once again set out to the south.

"Walking, walking, watching, watching," the brothers came to a ridge north of the land now known as Hedangna. They stopped to catch their breath and to rest. They finished up the last of their roasted barley and then climbed a tree to get a better view. They saw a deep-blue lake. This lake was so blue and so big and so surprising to see in the middle of the dense, dark forest that Minaba and Sepa were afraid. They looked closely at the water. In the center, they saw the wooden bowl and walking stick slowly spinning. The land surrounding the lake sloped more gradually than did any land they had yet seen. It faced the rising sun. There were still huge boulders, bigger than a house, but not as many as in the north. It seemed like good land. The brothers looked back at the lake, with the stick and the bowl circling in the center. They then climbed down from the tree and knelt by the spring that fed the lake. Each drank deeply from the cold, clear water. Together, Minaba and Sepa swore that as long as they lived, as long as their offspring lived, this land would be their home.

◊

As long as I can remember, I have longed to touch some sacred essence I did not have words for, something I knew only by its presence and mostly by its absence. This longing led me from my home; yet ultimately, I believed, it would lead me home, bring me to a place in the landscape—a place outside myself—where, like Minaba and Sepa, I would want to stay. And so I set out on my journey, traveling, as so many before me, to the Himalayas, and led, like the two brothers, by a longing for home, like them watching for signs that I was on the right path.

Thin Places explores what I encountered on my travels. Most broadly, it asks what it means to come home to a place in a culture without deep ties to the natural world. I asked this question during my fieldwork in Nepal; I asked it more urgently after I returned to the United States and a dissolving marriage. I found that truths that were easy to study as an anthropologist were more difficult to embrace as a way of life. I realized that I could understand the ways that others relate to the environments in which they live only by reflecting on my own relationship with the environment and that I could best understand that relationship by considering my relationships with others, particularly those to whom I was closest. I came to see that both sets of relations, with the earth and with others, required similar efforts and challenges and that I could really understand those challenges only by exploring them in my own life and choices.

I experienced the sense of connection I was seeking in those moments when the veil between worlds lifted, in those thin places where I could feel the presence of the divine. I came to understand that Minaba and Sepa's story about finding a home was really a story of pilgrimage, a journey that involves a stripping away to reveal the sacred essence of life. And I discovered that the ability to discover the sacred at home—living life as a pilgrimage—is what can turn any place into a home where it is worth swearing to stay.

ACKNOWLEDGMENTS

This book has been many years in the writing. I cannot begin to do justice here for the support, encouragement, and guidance I have received on what has been a much longer journey than I ever imagined it would be.

The research on which this book is based was made possible in the most immediate sense by the generous financial support from the Department of Anthropology at Harvard University; the Harvard Institute for International Development; a Fulbright-Hays Fellowship for Doctoral Research; the Joint Committee on South Asia of the American Council of Learned Societies and the Social Science Research Council, with funds provided by the Andrew W. Mellon Foundation and the Ford Foundation; the Hunt Postdoctoral Fellowship, administered by the Wenner-Gren Foundation; and the generosity of my grandmother Calvert Truxtun Holloway.

My greatest debt is for all of those in Hedangna, some of whose names I changed, who welcomed me into their community and their homes. I thank Dhanmaya and Dilli and their two eldest daughters, Kumari and Himali, for their generosity and their openness and, especially, Himali for her friendship. Both Chute Rai and Baiseti Thuma gently guided me toward the stories and rituals they believed most important about life in Hedangna. And I thank Raj Kumar for his good company on our travels around the village, his patience and persistence in helping me with my research, and mostly for his friendship.

The staff of The Mountain Institute provided a welcome intellectual and physical home during my stays in Kathmandu and, on my last trip to Hedangna, in Khandbari. I enjoyed numerous conversations over the years with Khagendra Sangam, Ang Rita Sherpa, Chandi Chapagain, Tashi Lama, Bob Davis and Gabriel Campbell. I thank them all for their support and friendship.

Thanks go to other foreigners whose stay in Nepal overlapped mine, especially Carole McGrananan, Claire Hefferen, and Kay Norton. A permaculture course led by Chris Evans and a trip through Jajrakot helped shape my understanding of the connections between culture and agriculture and enabled me to see and understand more about life in Hedangna than I would otherwise have seen.

At Harvard, I am grateful to the members of my committee: my adviser, Sally Falk Moore, for her intellectual rigor; William Fisher for his guidance and support at Columbia and Harvard and in Nepal; and especially, Pauline Peters. Pauline's intellect and integrity have given me an appreciation of what scholarship can be, and her encouragement helped me find the confidence to find my own way. A writing course with Verlyn Klinkenborg during my last semester helped me begin to envision the scope of this book. I especially thank Verlyn for his encouragement over the years and his admonishment to pay close attention to each word on the page.

Harvard would have been a very lonely place without the friendships of Lida Junghans and Julie Goldman. There is no way to express the gratitude I have for all that we shared and for all that their friendships have offered me then and since.

Madeleine Houston has offered me so much since that first glass of apple juice in Jawalakhel so many years ago. I am especially grateful to her for the times at Lake Ozette and for helping me find a way to bridge the worlds of Nepal and the United States. And I thank Madeleine and my sister, Sally Armbrecht, for welcoming me when I most needed a place to go.

I am very grateful for the men and women whom I came to know in the communities that my former husband, whom I call Brian in the pages that follow, and I visited on our trip across the United States. Although this is not the book they imagined we would write, I hope, in some way, that it contributes to their efforts to save the places that matter most.

A writing workshop with Terry Tempest Williams took this manuscript in a new direction, and her friendship over the years has helped me find the courage to follow it. I am also deeply grateful for the encouragement of the late Dana Meadows and, in particular, for her reassurance that my story was a lens into the larger world and that telling that story was crucial to what I wanted to say.

This manuscript has benefited enormously from the insights of those who have read it over the years. Molly Absolon, Terrence Youk, Lida Junghans, Madeleine Houston, Anne Davenport, and Verlyn Klinkenborg offered valuable suggestions on particular chapters that helped clarify the narrative in important ways. Camilla Rockwell, Robin Duscher, Carole McGranahan, Robert Buchanan, and my mother, Calvert Armbrecht, each read the entire manuscript at critical moments in its development and helped me articulate more clearly what I was trying to say. Kate Botham has been incredible, playing every role from cheerleader to loving critic to childcare provider. My debts to her are enormous. Woden Teachout appeared out of nowhere and has been the perfect reader and friend. This book is clearer, more concise, and more fully developed because of her input and her support.

Thanks also to Anne Routon for her belief and enthusiasm in this book from the outset. Her insights and those of the anonymous reviewers helped me develop and refine the manuscript more fully. I thank Irene Pavitt for her careful and thoughtful editing.

Then there are those who did not read the manuscript but provided me with what I needed in order to write. I thank Kate Collett for watching my daughter and for her friendship during a lonely winter and beyond; Mark Johnson for being such a strong, steady presence, insightful, clear, and kind during a time when that is just what I needed; and Nora Paley for helping me bring what I experienced in Nepal into my body. I also thank the Cobb Hill Cohousing community for their friendship and for understanding when I disappeared for days from community life, in particular Edie Farwell, Jay Mead, Matt Starr, Beth Sawin, Molly Shaw, and Lorie Loeb. I am deeply grateful to the circle brought together by Authentic Movement and Jan Sandman: Amy Goodman, Louise Low, Emily Medley, Kristina Triplat, and Kerrie Workman.

I thank Rosemary Gladstar for creating a place where I could begin to find my way home from Hedangna while keeping alive all that I encountered, and for her remarkable ability to nourish the seeds in everyone she comes to know.

This is not a story that Brian would have chosen to tell and were he to tell it, his account would be much different from my own. That said, I am grateful for the many experiences that we shared over the years and for his insights and perspectives, which have informed and broadened my own.

I thank my parents, Calvert and Ted Armbrecht, for their curiosity and interest in my work, wherever it has taken me. They made the long journey to Hedangna, sleeping for a night in my bamboo-slat room before continuing our trek to the Makalu Base Camp. Their love and support has been unconditional.

My brother-in-law, Peter Absolon, was killed just after I completed this manuscript. I thank Pete especially for showing me, in his life and his death, how to live as fully as I possibly can.

I am grateful to my daughter, whom I call Avery, and to my son, Bryce, for so much, especially, perhaps, for offering the ongoing opportunity to find the sacred at home. Finally, I thank Terrence Youk, with whom I have found that clearing I longed for.

PART ONE
departure

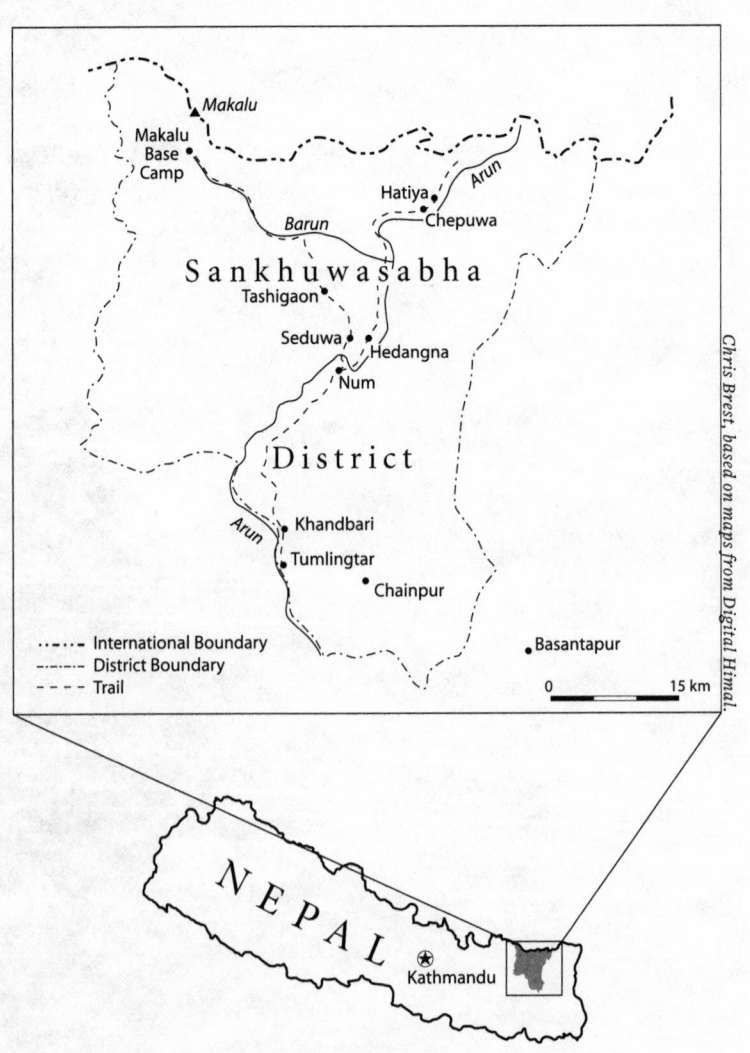

Makalu

Makalu
Base
Camp

Barun

Hatiya

Arun

Chepuwa

S a n k h u w a s a b h a

Tashigaon

Seduwa

Hedangna

Num

D i s t r i c t

Arun

Khandbari

Tumlingtar

Chainpur

Basantapur

····· International Boundary
─·─· District Boundary
─ ─ ─ Trail

0   15 km

N E P A L

Kathmandu

1

GROWING RICE

July 1992

HEDANGNA, NEPAL

I stood ankle deep in mud, bent at the waist, and pressed pale green rice seedlings into the gritty mud. The mud was thick and brown and felt cool around my feet. My damp lungi clung to my thighs, and my shawl was draped over my head to keep out the morning rain. I finished planting the seedlings I held in my hand and waded across the terraced field to get another handful from a bamboo basket on the edge of the terrace. I waded back to my place, stubbing my bare toes on small rocks as I went; bent down; and again pressed the seedlings into the mud. I paid attention to the work at hand, my mind lulled by the rhythm of planting, by the voices of the dark-haired women working at my side, by the slosh of mud and the rippling of water.

The soil in the field where we worked was dark and fertile. The survival of these women and their families and of this village in eastern Nepal depended on this soil and what it produced, especially rice. The rice season began in May with the *bali puja* (crop ritual). Villagers built altars for the ancestors and laid out offerings of chicken and *jad* (millet beer), eggs and rice: gifts of food and drink in exchange for rain. After the ceremonies, the rains began and the streams flowed, bringing water to the cracked earth. Men woke before dawn, filled a wooden cask with beer, and drove their teams of oxen down the hill to plow their terraced fields. I woke some time later to the sound of rain on bamboo leaves outside my window and the shouts of the men as they steered the oxen through the mud, plowing the soil in preparation for planting the rice. I tried to plow once and was surprised by the feel of the earth through the shaft of wood, surprised by the intimacy of the connection between the men and the soil, even though they were not digging through the gritty dirt and sharp stones with their fingers, as the women did when they came through with the seedlings after the fields had been plowed.

The structures in the village, the shrines and houses and fences and fields, were built to be lived in and worked on. If they were not used, the forest quickly crept back in. Wind and rain tore down altars made of bamboo and saplings several days after a ritual. Fallow fields became overgrown with weeds and ferns after a year or two and were reclaimed by the jungle after five. The stone-and-mud houses were dank and cold until they had been lived in, until cooking fires had dried out the chill. They cracked and crumbled if fresh mud mixed with cow dung was not rubbed over the walls on the mornings of the full moon and the new moon each month. The thatched roofs became matted and began to leak if they were not replaced every year; the moisture then loosened the rocks from the mud, and the walls began to fall. Abandoned houses were inhabited by pigs and goats after three or four years and soon became piles of stones.

If Hedangna were deserted for thirty or forty years, the moss-covered rocks and the terraces would be the only remains, but the stones of the terraces would have eroded, the walls collapsed, and the fields become covered with trees. The land continued to be habitable only because of the ceaseless work that kept it that way.

The weather and seasons as much as the villagers' own needs and objectives shaped the rhythm and pace of work. They lacked the tools needed to submit the land to their will. Instead, they moved across it slowly, by hand, feeling their way for openings, for places where the plow would move smoothly and the seedling would stand, turning from places where the earth would not give. This knowledge was not broad or transferable; the villagers became less sure of things when they found themselves at a higher or a lower elevation. What they knew was rooted in the land that sustained them.

◊

I had come to Hedangna to conduct research for a doctorate in social anthropology, to understand how the Yamphu Rai got rights to the land and how they held onto that land. I wanted to understand their perspectives on the area's recent designation as a conservation area, a buffer zone to a new national park a few days' walk up the ridge. I was also there because I wanted to discover how to live more simply and more lightly on the earth. I assumed that most problems in America stemmed from our disconnection from the land, economically and culturally. In Hedangna, the impact of the villagers' ways of living on other people and the earth were immediately and concretely visible. Trees cut this year meant walking farther for firewood next year. A dispute with a neighbor meant one less person to turn to for labor or help. I wanted to understand what difference those close connections made in how they cared for one another and the land. And I was here because I believed that indigenous people had wisdom about how to live on the earth that we had lost. I wanted to understand whether that was true and, if so, what that wisdom was.

Hedangna is the largest village in what has come to be known as the Makalu-Barun region of northeastern Nepal, which begins to the east of Mount Everest and extends to the Arun River, one of the world's deepest river gorges. Hedangna, the largest Yamphu Rai village in Nepal, is a community of around 273 households spread out at an elevation ranging from 2,700 to 5,800 feet along a gradual, east-sloping ridge above the upper banks of the Arun River. Distant mountains encircle the village, and when you are within these mountains, it is difficult to imagine being anywhere else.

The land is remote, even by Nepalese standards. Lines of communication between Hedangna and the outside world are tenuous. Some families have radios. Month-old newspapers occasionally arrive at the Nepal Bank at the southern edge of Hedangna, but few villagers bother to read them. It is a hard five-day walk to the nearest road and then an eighteen-hour bus ride to Kathmandu, the political and economic capital of Nepal. (Villagers measure distance in terms of time taken rather than distance covered.) Since the early 1980s, those with money for the airfare had walked for two days to an airstrip in Tumlingtar, a hot sleepy village along the Arun River several hours south of Khandbari, the district capital of Sankhuwasabha. But few villagers have the money to fly.

Each January, one or two members of each household walked for a day and a half south to Khandbari or for a week farther south to Basantapur to purchase their family's yearly supply of salt and kerosene with money earned from selling rice to Bhotiyas from the north or to civil servants posted in the region; others went to the district court in Chainpur, a half day's walk east of Khandbari, several times a year to plead their case in a land dispute. Brothers or sons who had emigrated to India sometimes returned for a visit, to check on their land or to bring back money and Indian cloth, and families with a son studying in Biratnagar or Kathmandu occasionally received a letter asking for more money.

For the people who stayed in Hedangna, the world beyond this valley was distant—beyond what they could imagine. An older woman once described the difference between their lives and mine. "You can see a plane, can get up close and touch it and climb inside," she said. "We can only see its silvery bottom, high overhead, soaring through the sky."

In the evenings, after a plate of rice or millet, men and women sat cross-legged on straw mats laid out on the mud floor, staring into the coals and talking about how many loads of firewood they had left to carry, whether they would plant cardamom in their forest, what would happen when the land survey came. Then they crawled under blankets woven from the wool of sheep grazed on pastures north of Hedangna and fell asleep to the dull roar of the Arun, a mile to the east.

In the mornings, Devimaya, a strikingly beautiful woman in her twenties who became my closest friend in the village, often asked what I had dreamed about the previous night. Sometimes I had dreamed about land disputes or the names on a genealogy. More often, I had been caught in a traffic jam in the polluted streets of Kathmandu or had been racing through a crowded airport to catch my flight home. When I asked her, she would sigh and say with exasperation, "If I carried firewood, I dream about carrying it all night long. If I planted rice all that day, I spend the night planting rice. I have to do this work all day—at least I could do something else in my dreams."

◊

I was working in the rice field of Raj Kumar's family. Raj was a few years younger than I and sometimes helped me with my research. Like most schoolteachers in the village, who spent their days behind a desk, he was a mix of the rough physical strength of a villager and the smooth polish of a civil servant—qualities that were more or less accentuated, depending on the people he was with and the clothes he was wearing. He wore wire-rim glasses and store-bought clothes, and he did not go barefoot. Each morning, he showed up at the water hole carrying his towels and toothbrush, like me. Everyone else brought jugs to fill with water, lungis to wash. They used rocks to scrub their feet. He, like me, used soap.

I had met Raj Kumar the first week I was in Hedangna. His younger brother came to my room the second morning after I arrived to ask if I would tutor him in English. As he left, he told me that his brother had been to Kathmandu and that I should meet him. I went to Raj's house that afternoon and soon began to go there every day before dinner. Raj had spent two years in Kathmandu, studying for his Master of Education. Then, in his late twenties, he had had to return home because he had no more money and could not get a job and because his parents wanted him to come back. They arranged for him to marry one of the few Rai women in the region who had passed the high-school examination, because, as Devimaya said, someone who had studied in Kathmandu had to marry someone who could read. Raj, his wife, and their eighteen-month-old son lived in a small room next to his parent's house, and Raj taught math in the village high school.

Raj was the first Yamphu who understood what it meant to conduct re-search, and he was the first villager with whom I really felt comfortable. I took copies of *Newsweek* to his house in the evenings before dinner. When I showed pictures in the magazines to women in the village, they slowly rotated the im-ages, trying to find something they could recognize. Raj understood the pho-tographs right away. He had seen cars and buses, tall buildings and jets. The pictures expressed something known, a window into a world with which he was familiar and in which he could immediately get his bearings. We looked at the photos and talked. He explained Nepali words to me, corrected my pro-nunciation. I explained world politics to him.

Raj depended on me for remembering the parts of himself that longed for something else, that did not want to be in Hedangna, did not want to be mar-ried and responsible for a small child. I reminded him of the dreams he had had as a young student in Kathmandu during the revolution of 1989. I could understand why he longed for the freedom and independence he had expe-rienced during that time. That same longing had brought me to Hedangna. He could understand why I had chosen to leave my own home. Yet he could never grasp why, of all the places in the world I could have gone, I had chosen to spend two years of my life in a place he longed to leave.

In many respects, Raj believed more in the promise of the West than did I. He had tasted just enough of the modern world to think it was all good. Once he looked at his pen, which was identical to mine, except in color, and told me that my pen was better than his. Another time, his wife offered me a snack, salty bread sticks bought in a local tea shop and, like anything purchased in the village, considered a real treat. Raj waved his hand in disgust, saying that this snack was not nearly as good as the snacks and tea that I drank in my own room. I could look at my things and know what had quality and what did not. He could compare them only with his own. Measured in this way, his possessions inevitably came up short.

◊

Raj was plowing when I walked down to the field, and as he greeted me, he looked at the mud on his clothes and at his bare feet. "You don't plow like this, in your country, do you?" he asked. "You use machines, and don't have to muck around in the mud. Isn't that right?"

I was as embarrassed by his questions as he was by the mud on his arms and hands. I answered as I always did when asked questions like this: yes, we farmed differently, but that did not necessarily mean that what we did was better. I said something about soil erosion from mechanized farming and then told him that I had come to help with the planting. He nodded and went back to plowing.

I took off my sandals, which were too heavy for the ankle-deep mud; climbed over the edge of the field into the muddy water; and asked his mother how I could help. She handed me a clump of seedlings and told me to plant next to Altasing's wife.

I waded through the mud and began to plant as quickly and as carefully as I could. Altasing's wife greeted me and immediately started to ask questions: What were Ganesh and Jaisita, in whose home I lived, doing that day? Did they eat rice or millet in their house (a question that was more an inquiry about their economic status—only the most wealthy could eat rice every day—than about taste)? What was the English word for "penis"? As soon as I answered one question, she asked another. In between questions, she told me to plant closer to the edge, or Raj's mother would yell at me.

We worked steadily for some time, interrupted every now and then by Raj's mother coming to see how we were doing. She shouted at her daughter-in-law to spread the mud around the terrace more thoroughly. She ordered her husband to get to work. And she yelled at me to plant the seedlings closer to the edge.

Raj's mother scared me. She looked like my childhood image of a witch: her dark skin was dry and wrinkled; strands of black hair streaked with silver stuck out around her face before being pulled into a tight braid that hung down her back. Her green shirt was torn at the sides, and her dirty lungi was bunched up around her waist. Whenever I visited Raj Kumar's house to speak with him, his mother offered me *jad* only after Raj had insisted. This beer was thick and slightly sour. It was the only beverage most families drank. Offering it to guests was the hospitable thing to do. As Raj's mother handed me a bowl, she always commented that all I did was talk and write; that I did not have to "work," as they did; and that I had not done anything to deserve this beer. I always accepted her words and the beer without comment. She was right. My work was a luxury to the villagers, especially the women, who hardly ever had a chance to sit around and talk. There was nothing for me to say. This was the first time I had gone to help in her family's field, and I wanted to prove that I was able to do her kind of work.

After what seemed to be a long time, Raj's mother called us over to the edge of the field, where she had prepared some *jad*. We rinsed our hands in an irrigation ditch that brought water into the fields and gathered in a small circle on a huge boulder. The women talked about how many terraces still had to be planted. They talked about who was planting their fields and where they would work the following day. Raj's mother passed me a bowl of *jad,* along with everyone else. She urged me to drink it so she could fill it again.

When there was a lull in the conversation, she started to talk about how hard and how long I had worked. If I did not already have a husband, she said, she would marry me off to Deuman, her youngest son. The women laughed. Grateful to be included in her jokes, I replied that yes, yes I would of course marry her son, even though I was at least twelve years older than he was. I could have two husbands, I said. The conversation shifted to Brian, my husband, and what he was doing in America. They asked again why he was not with me and whether, when he came back, he would bring the photos that he had taken when he was here before.

The women rolled tobacco in leaves, rubbed a rock against a piece of flint to create a spark, and puffed on their homemade cigarettes. After we finished our *jad,* Raj's mother handed a clump of the leftover fermented millet used to mix the beer to her daughter-in-law and then one to me. Raj, who had been silent until now, sharply told her that it was disgusting to use her bare hands, that of course I did not want to eat the millet. She started to pull her hand back. "No, no," I said. "I like that part of the *jad*; it isn't disgusting to me, at all." She looked at Raj, uncertain who to believe. I again insisted, and she reached over and handed me a fistful of millet.

◊

In the evenings when there were batteries that worked, Devimaya would turn on the radio to hear the news. The loud voices, music, and static seemed out of place in the dark room where we quietly watched the dying coals. The smell of smoke hung in the air. There were reports about royal ceremonies and political struggles, events that seemed so far away and unreal that it never seemed worth trying to follow what was happening.

I listened instead to the advertisements for soap and clothing, jangly music and cheerful messages coming from a world where I imagined everything to be sparkling and clean. Sitting on a hand-woven straw mat on the mud floor near the fire pit where our meal of rice and steamed nettles had been cooked in blackened pots, I pictured the recording studio in Kathmandu or Biratnagar where the clips were taped: rooms crowded with young men and young women, freshly scrubbed and dressed in pressed clothing, sitting in wooden chairs at a shiny table where they sipped cups of sweet tea. Life was easy there: water poured from taps, lights turned on with the flick of a switch, rice and lentils could be bought at a store, and twenty miles could be covered without breaking into a sweat.

As I listened, the room where I sat suddenly seemed darker, the layers of soot and dust covering the beams seemed thicker. I noticed that the children were barefoot, their clothes ragged and worn. What had been a normal evening in Hedangna after a normal day's work suddenly felt like one more dull and tedious evening stretching into a string of dull and tedious evenings. Chores seemed more endless, and the cold and the wet cut in more sharply.

I often joined the women in the fields, helping with digging and planting and cutting and carrying, doing whatever I could to create something for us to share. Although I was slower and clumsier than they, they welcomed the free labor and the novelty of having me around. During breaks in the work, when we gathered on a rock or under a tree, the women, old and young, would reach for my hands and rub their fingers slowly across my skin. They would turn my hands over and feel the palm, pull the fingers up to their eyes, and comment about how smooth and white they were. Then they would hold up their own hands and feet, which were tough and dark, next to mine. They looked at one another and shook their heads. They lived by their hands, they would say, and I lived by my head.

The women in Hedangna want skin like mine. They want some padding in their lives, want to be able to stay inside for a while and let their bodies become smooth and white and soft. I want skin like theirs, dark calloused skin that lets them walk through their lives barefoot, enduring, not avoiding, the sharp pain encountered on the way.

I was raised in a world where what was valued was what I could know with my mind. I was educated away from my home, taught that there was more to be gained by moving forward than by staying put. I left my home to understand what it took to stay at home, went halfway around the world because I wanted to learn what it meant to live with my hands and my feet and my heart—to remember what these women's ways of living have never let them forget.

2

SEEDS

December 1991

When I first arrived in Hedangna, I stayed in a tiny room on the porch above the office of the Small Farmer's Development Bank in Gadi. Gadi is the Chetri village on the southern side of what is broadly called Hedangna where the bank, the post office, the police post, and a few shops that sell tea, biscuits, kerosene, Indian-made cloth, cigarettes, matches, and other sundries are located. I hoped to live with a Yamphu Rai family, but it would take time to find someone with enough rice to feed an extra person. I stayed in Hedangna-Gadi while I waited.

I ate with a Chetri family that lived across the path from the bank and was often joined by two Nepali civil servants: tall, thin men placed for two years in what was considered the hardship post of Hedangna. Although Hedangna is remote, they were happy to be here because the hardship designation meant that they earned twice as much pay. During meals, they talked about news from Khandbari, the district center, or Biratnagar, a large city in southern

Nepal, and about where they hoped to be posted next. They asked me about the United States and why I was here. Mostly, they joined the Chetri woman who served our food in her efforts to dissuade me from moving to the Rai village just up the trail.

She spooned the dal onto my plate, shook her head, and told me that the Rai drank *raksi* (millet wine) and *jad*. The men nodded and added disapprovingly that the Yamphu mixed their food together on one plate, unlike the Chetri, who served their rice, dal, and vegetables on tin plates with dividers. She said that the Yamphu were ignorant, backward, and dirty. Why did I want to live in their village when I did not have to?

The Muluki Ain (National Legal Code) of 1854 formalized caste differences in Nepal and spelled out everything from which castes could eat what food and with whom to which castes held their lands under a particular system of tenure. The Kiranti, the broad ethnic category to which both the Rai and Limbu groups belonged, was listed in the Muluki Ain as one of five caste groups in Nepal. Along with the Sherpa and other Tibeto-Burman groups, the Kiranti were considered members of the drinking caste. The Chetri, who were in a higher caste referred to as the "wearers of the sacred cord," were far more preoccupied with pointing out caste distinctions than were the Yamphu, perhaps because the Chetri had a status they needed to maintain and the Yamphu did not. I was trying to be culturally sensitive and understanding of differences, but this practice of putting people into categories that then determined their value made me uncomfortable. After two weeks in Gadi, I was relieved to move in with Ganesh and Jaisita and their four children, a Yamphu family in the upper village.

When I went to Hedangna, I thought that the greatest cultural distance I would have to cover was the one between the Kathmandu Valley and Hedangna, if not that between the United States and Nepal, naively assuming that the move from Gadi to the Rai village would entail just walking for thirty minutes up a dirt path, never expecting that the distance between two parts of the same village could feel as great as that between my home in Massachusetts and Hedangna.

The area is remote physically; it is also remote intellectually. Before the 1950s, the prime ministers allowed only Brahmans to attend school. Anyone else caught studying is said to have had his hands cut off. Foreigners were prohibited from entering Nepal until the early 1950s, and since areas along trading routes with Tibet were closed when the Chinese invaded Tibet in 1959, Hedangna was off-limits to any outsiders until the late 1980s. By the time I arrived in the early 1990s, villagers in Hedangna had heard of trekking, and most had seen foreigners when they went to Khandbari to stock up on supplies, but few had ever spoken to them. One other researcher, Roland, a linguist from Denmark, had lived in Hedangna for six weeks the previous year. Otherwise, hardly any non-Nepalese ever passed through the village.

Except for Ganesh, the family did not speak much Nepali. They hardly spoke at all, especially to me. They commented on almost everything I said or did in the Yamphu language, which is not at all related to Nepali. At first, I made an effort to learn Yamphu. I carefully wrote down words and phrases. I practiced greeting villagers in Yamphu and learned enough to talk about their work and their other activities before falling back on Nepali to continue the conversation. But I soon realized that no matter how hard I tried, I would never be conversant in the local language and could never learn two languages and still complete my research in the eighteen months I had assured Brian it would take. And so, although I was able to converse quite easily in Nepali, language differences always marked a distance between the villagers and me.

Ganesh was one of the wealthiest men in the village, which simply meant that he had enough land to produce a surplus of rice that he could sell to Bhotiya traders from the north. The wide and airy porch, with the feel of a veranda in a Latin American hacienda, and tin (rather than thatch) roof of his house were the only visible signs of this wealth. Otherwise, his house was like every other house in the village: made of stone and mud, with one ground-floor room connected by a ladder to a dark second-story room that was used primarily for food storage.

My room in Ganesh's house, a space just large enough for two sleeping bags to be spread out, was one of the nicest I had in Nepal. It was on the northeastern corner of the second floor of the porch and was enclosed by thin strips of bamboo placed loosely together, so that light and air filtered through the slats. An opening in the bamboo served as a window that looked over the terraced fields and across the valley to the snow-capped peaks beyond. Because of the airiness of the room, the only way I could be comfortable in winter was to dress in my warmest clothes and sit in my down sleeping bag. But unlike the rooms inside the house, whose doors admitted the only natural light, my room was sheltered from the weather but not separate from it. It had a combination of openness and seclusion that I loved.

During my first week in the village, neighbors would climb the narrow ladder to the second-floor porch and enter my tiny room without knocking. They sat cross-legged on the straw mats, silently looking at my sleeping bag, my books, my notebooks, my camera. While packing for my stay in Hedangna, I had been proud that I was bringing only what I could fit into a backpack and a metal trunk. Now in this village, where most people did not have enough money to buy kerosene to light a lamp at night, my relative affluence felt obscene.

In the morning when I went to the water hole to wash, dark-haired women with gold nose rings hanging past their lips picked up my soap. They reached for my toothbrush, pointed to my toothpaste. They commented on my big rubber sandals, my skirt, and my hairbrush and laughed when I said *bat* instead of *baT*.

A few days after I moved in with Ganesh and Jaisita, I went to the forest to gather wild bamboo shoots with two other young women. As I tried to keep up with them as we scrambled through the steep rocky forest, I slipped and fell down hard on my tailbone. The two women burst out laughing, clutching their stomachs because they were laughing so hard. I laughed, too, to keep from crying. The next day at the water hole, a woman I did not recognize patted her tailbone and laughingly repeated the story in Yamphu to the crowd of women who were filling their jugs with water.

When the Yamphu meet someone new, they introduce themselves first with their clan and then with their *tsawa*, a Yamphu concept that refers to the spring from which their ancestors first drank on arriving in the village. This information immediately allows them to insert the individual into a social map that is attached to a physical and historical landscape.

All they knew about me was my name, which did not mean anything to them. They called me simply "little sister" or "older sister," depending on the age of the speaker. They did not understand why I had come to Hedangna alone or why I did not seem to have any friends. They could not understand why I unafraid to sleep on my own and why in the world I would want to.

In this context, the emphasis on caste began to make some sense. With caste, it doesn't matter whether you know someone personally; you know the category in which he or she belongs, and that can guide your intentions. Like any category, caste is confining, and yet it also provides a basis for knowing whether and what to trust.

◊

Shortly after I moved in with Ganesh and Jaisita, Ganesh's younger sister, Padma, came to my room with her cousin Devimaya. Both were in their early twenties, but, unlike most women their age—who had dropped out of school, married, and borne several children—Devimaya and Padma were studying for their high school–leaving certificate. They lived at home, so they had more control over their lives than they would if they were daughters-in-law. They sat on the floor of my room and, as their eyes took in my sleeping bag and stove, my camera and books and photographs, asked if I would tutor them in English, as I was tutoring Deuman, Raj Kumar's younger brother, who was in their class. Grateful to meet some women with whom I had something in common, I readily agreed.

Unlike Deuman, who came to my room early each morning, Devimaya and Padma never actually came for any tutoring. Although they were studying, they were women and, exam or not, women in Hedangna had to work. And so it was not until almost a month later that I bumped into Devimaya at the water spigot in the upper village, where she was washing a dark blue lungi. She greeted me and asked me to her home for *jad*. I followed her as she quickly walked up the stone wall, passed through several packed-dirt courtyards, and ducked into a cool, dark house. She pointed to a straw mat next to the fire on which I should sit and sat down on another mat closer to the fire. She blew on the coals of the fire and heated a pot of water. While mixing clumps of fermented millet into the warmed water, she asked if white women also bled each month and, if so, what did they do. And she had read in school that people—she forgot what they were called—had been to the moon. Was that really true? The questions kept coming as she pounded red chilies for chutney, poured the slightly sour, earth-colored beer into a small brass bowl, and then sat back on her heels to watch me drink and hear my replies.

◊

A few days after settling in at Ganesh's house, I brought out packages of seeds—carrots and basil, spinach and tomatoes—that I had bought in Kathmandu. I hesitantly asked Ganesh whether he had any unused land where I could plant some vegetables and herbs. He scoffed at my request, not, as I had expected, because he had no spare land, but because, he said, it was not worth the trouble. The chickens would eat the seeds. The goats would eat whatever seedlings made it past the chickens, and the neighbors would "steal" whatever made it far enough to be edible. I thought that these were just excuses for not having a garden himself and insisted that at least I wanted to try.

Ganesh finally pointed to a bit of rocky land near the side of his house, in the shade, and then watched from the porch as I went to work clearing away the weeds and digging up the soil. He watched while I planted the seeds and covered them with leaves and hay for mulch. He did not say a word. As I worked, I thought about how good it would be to bring carrots and basil to Jaisita to cook. Neighbors passing by on the trail stopped to watch what I was doing. They shook their heads and repeated what Ganesh had told me. I tried to ignore them and went to get water from the water hole. I made a makeshift fence out of brambles gathered by the stream and then stood back and proudly surveyed my work. Every day for a week, I carried water from the stream in an urn on my back. I repaired the fence where chickens had knocked it down. They knocked it down again. After the third time, I did not bother to repair it. It was time to begin my research.

3

CONSERVING THE LAND

June 1989

The summer after my second year in graduate school, I had gone to Nepal to examine indigenous systems of forest management in the Makalu-Barun region, to the east of Mount Everest, Nepal's most recent national park and conservation area. At a time when other national parks were excluding villagers, the Makalu-Barun Conservation Project promised to work with them; while personnel at many international development projects believed that encouraging development and conserving the environment were mutually exclusive, staff at the Makalu-Barun project believed that one could not be considered without the other. Nepalese researchers had been hired to conduct research for two years before programs were even introduced, in itself a novel idea in development, and they were asking tough questions about how to integrate the often conflicting needs of culture, development, and conservation. I assumed that they would continue to ask those questions even after shifting from research to implementation. The project seemed smart, and the people employed were smart. I wanted to do research that made a difference and was interested in finding a way to have my dissertation connect to the project in some way. My research that summer was a step in that direction.

Ten days after arriving in Kathmandu, I flew to Lukla, in the Khumbu Valley, with Lok Bahadur, my research assistant, to gather information on grazing use and on a small so-called conservation area created by a Sherpa lama and identified by Makalu-Barun Conservation Project staff, who thought it might be an interesting example of an indigenous system of conservation.

In the United States at the time, the model that most influenced the land-conservation movement focused on protecting wild places, where, it was assumed, people did not live. This model, which created its own problems in the United States, was devastating when exported to developing countries, where land that the governments claimed to be empty usually was nothing of the sort. Sometimes, areas that governments designated as national parks actually were settled; other times, they were traditional sites for grazing or sources of fuel wood and other nontimber forest products that communities depended on to survive. After disastrous results from relocating whole villages or from forbidding the villagers to use the now enclosed resources (the border of Sagarmatha National Park was denuded after the cutting of trees inside the park was prohibited), international conservationists came to recognize the need to work with those who lived in and utilized the resources in areas targeted for conservation. A key strategy in this new effort to create projects that had legitimacy on a local level and that actually accomplished their objective of protecting resources was finding indigenous management practices, particularly those by which villagers seemed to impose their own restrictions on resource use. I was thrilled to have discovered such a system to explore in more detail.

Brian, who had stayed behind in Cambridge, Massachusetts, for the summer, planned to meet me two-thirds of the way through my trip at a tea shop in a village southeast of Lukla that we had selected from a map. We confirmed this plan before I left Kathmandu to begin my research. It was the height of the monsoon season, so flights from Kathmandu to Lukla were undependable. I would have no way of knowing whether Brian had made it to Lukla, no way of knowing whether he had even made it to Kathmandu. We had made only a vague contingency plan, not willing to consider the possibility that Brian might go to all the trouble and expense of traveling to Nepal for a three-week visit, only for us to miss each other. We trusted fate to take care of what we overlooked.

Lok, a porter, and I spent three rainy weeks traveling to remote villages, where we tried to talk to inhabitants suspicious of both the national park and our questions, before reaching the spot where I planned to meet Brian. The shack, which I had been told was a tea shop where we could stay, was closed, the door boarded shut. The so-called village was nothing more than this and another hut in an otherwise deserted grazing pasture a two- or three-hour walk from any village.

We stayed in the grazing hut and waited. By the end of the first morning, I had finished copying my notes from the interviews we had conducted in the villages we had visited. No one was around, so there was no new information to gather. I watched the sky, heavy with thick rain clouds, dark and gray even when it was not raining, and listened for sounds of small planes. I asked every Nepali who passed by, maybe four a day, whether the plane had flown to Lukla and whether they had seen a tall, thin foreigner heading this way. Day after day, the flights were cancelled. We ate two meals of boiled potatoes each day, watched the rain, and waited for time to pass.

We stayed as long as we could bear to, and then decided to head to the village with the conservation area, which lay a few hours to the west. I had been told that the forest behind a monastery had been set aside by the lama of the village. I tacked a note for Brian on the wooden door of the hut, telling him where we had gone and giving him instructions for how to join us. We then headed down the narrow trail, overgrown from lack of use, that led to the village. During the dry season, branches overhanging a trail are not a problem. During the monsoon season, those branches are covered with heat-seeking leeches. This trail was covered with overgrown branches, and every fifteen minutes we had to stop to remove leeches from our shoes and ankles, our arms and hair. Early on the trip, Lok had showed me how to tie a bundle of salt to the bottom of a stick and press the salt tip on leeches as I walked. The salt makes the leeches curl up and roll away from digging their heads into the skin, eliminating the need to stop every few minutes to pull them off. I thought that I was used to leeches after the past month of traveling in one of the wettest parts of the country in the wettest month of the year. But this trail was different. The salt stick was useless. I felt leeches crawling up my legs and digging into my skin, even when they were not.

Conditions did not improve much after we arrived in the village. Everyone seemed suspicious when we explained why we had come. Finally, an old woman reluctantly let us stay in her home, which consisted of one large room. We quickly discovered the reasons for the suspicion. The so-called local model of conservation was not at all what I had been led to believe before coming to the village. The lama was trying to claim land that was not his under the guise of conservation. He thought that we were on the national park's side and that our endorsement would legitimize his actions. The villagers thought that we were on the lama's side and would steal their land. Neither the lama nor the villagers believed our claim that we had come only to gather information in order to understand what was going on.

Early in the morning after our one full day in the community and a day before we planned to leave, a group of village elders came to the house where we were staying. They silently filed into the room and sat in a row against the far wall. Lok and I sat quietly, drinking our tea and watching as one of the oldest men in the group began to talk, calmly at first, less so as he went on. He described who used the forest and explained how it had been used in the past. His voice rising, he talked about the lama and his attempt to appropriate the villagers' land. Another man interrupted, and then another threatened "trouble" if we did not leave the village immediately. They waved their arms to emphasize their words. Others joined in, and their voices became louder. Lok nodded and calmly repeated our reasons for being there, our objectives, and our affiliations. I sat quietly at his side, taking more from the anger in the men's faces and gestures than I could from their words. Lok told them that the park would bring bridges and water pipes. He promised that no one would take their land. He repeated that we were just gathering information.

Lok thought that he had appeased the villagers enough to allow us to stay the night, as we had planned, but I wanted to leave immediately. I already lacked confidence in my ability to conduct research; the villagers' anger made me doubt myself more. It also made me question the assumptions about conservation and indigenous knowledge that had brought me here in the first place. I did not see any reason to stay. And so we packed our bags and headed out of the village.

Just as we began the descent into the leech-covered brush, we saw Brian walking out of the ravine with a Nepali porter. Before even greeting him, I exclaimed that he should not be wearing shorts on a trail like this. In reply, he told me that there was a leech above my lip.

◊

Brian's plane had been cancelled for three days. On the last possible day he could still reach me, he had made it to Lukla, found a porter and guide, and begun the two-day hike. He had arrived at the designated meeting place earlier that day, carrying a cup of raspberries collected along the trail. When he discovered my note tacked to the grazing hut telling him to hike for three more hours, he ate the raspberries.

He looked at me incredulously when I said that we were going back down the leech-covered trail to spend the night in the abandoned hut on which I had tacked the note. I told him that I was sorry—really, really sorry—but we had no other choice. We could not return to the village, no matter how tired he was. There was nothing he could do, so he pulled on a pair of long pants, and we turned and set off down the trail.

We did not talk much on the return trip. Brian was exhausted. I was distracted by the fear that the villagers would chase us, as they had threatened to do. I wondered what I should have done differently, how the Makalu-Barun Conservation Project staff would have handled the tense situation. I worried about where we would stay that night and what we would eat. I worried about Brian and how he was responding to this sudden change in plans. I was used to being on my own and, by now, used to traveling in the monsoon. I also tended to put up with a high level of discomfort, accepting it as part of the experience rather than going to what seemed like the additional trouble of making conditions easier. Brian was more inclined to try to improve a situation, not just endure it. That night, I imagined, we would not have much choice either way.

We returned to the grazing hut to find that a cow had been killed by a wild cat that afternoon, an unusual event, and that a group of herders had gathered to celebrate in the hut where we hoped to stay. Dinner was taken care of: we ate a huge plate of meat. We then spent the night in the ten- by twelve-foot hut with fifteen or so Nepalese, crowded together under the leaking straw roof. I opened my umbrella and placed it over our heads, as much to keep out the rain as to give Brian and me a bit of privacy. It was not the meeting we had envisioned. But that, as much as anything, was what I loved about Nepal.

◊

Two weeks later, Brian and I flew from Kathmandu to Alaska. Brian was to continue to Boston, while I was to spend a week in Anchorage developing an alternative research project that would be easier on our marriage. But when the plane landed, I did not get off. Even with the never-ending rain, the difficult villagers, and the leeches between my toes, as I sat by smoky fires and listened to men and women sharing stories about their lives, I had known I was never going to do research in Alaska. Despite the difficulty of working in rural Nepal—indeed, because of it—I came alive there in a way I did not anywhere else. I had to return to Nepal to keep that self alive.

I did not admit this to Brian. I talked about the conservation project, the chance it provided to work with scholars who had similar interests, and the remarkable opportunity it offered to conduct research that had a use.

Brian did not think much of anthropology or academics, but he believed in conservation and was even more romantic about rural farm life than I. Both of us cared deeply about doing work that made a difference and about finding ways of living less destructively on the earth. Both of us believed that reconnecting people to the land was a crucial part of fixing what seemed wrong in our culture's relationship with the earth. And so Brian understood my reasons for wanting to return to Nepal. He believed in my project. He told me that we would figure things out and that everything would be fine. I believed him.

◊

I feared what would happen to my marriage if I returned to Nepal. I feared what would happen to me if I did not. Even now, I do not know how Brian felt or what he thought. Both of us thought that shared intellectual interests and values were what we needed to make love last. In our heads, it all made sense. Neither of us was conversant enough with the language of our hearts to hear what they had to say.

4

THE BOOKS

September 1991

Before I went to Nepal to begin eighteen months of fieldwork, Brian and I hosted a party at the Moosilauke Ravine Lodge to celebrate his thirtieth birthday and to bid me farewell. Brian and I had met in college at an outing club meeting. He asked me to marry him at Kala Patar, at 18,000 feet looking across to Mount Everest, and we went mountaineering in Bolivia for our honeymoon. A party at Mount Moosilauke in New Hampshire, where we visited again and again to hike and backcountry ski, seemed the perfect way to honor and renew our commitment to each other before I left for Nepal.

Friends and family from across the country gathered in time to hike to the summit and back before dinner and a square dance. As everyone finished eating, Brian stood up and presented me with a leather photograph album with my name and the date inscribed on the cover. He had enlarged photographs from trips we had made together—kayaking, backpacking, backcountry ski-ing. Interspersed with the photographs were cards sent by friends. The most important part of the book for me was a poem he had written: "Finally, a Poem":

When we first met, you were reading Henry James
and I was reading "Zen and the Art of Something."

You gave me a copy of "The Waste Land" and for me
that's what it was . . .

for you, Ann, it's about remembering
the flood in West Virginia, some unknown hollow, or the next peak . . .
What lurks that may be pure, that still remembers its own hometown.

There's a sacred place you never knew, and never forgot
that may lie deep in the West Virginia hollow or the Buddha's stupa
on a clear cold night as the chanting goes around.

You have not forgotten what I will never know.
A culture lost, but not forsaken.

To me, a sacred place is still that swimming hole off Route 118 or
anywhere on a summer night in New England
under a thick canopy of trees
or that flash of water when your bow is upstream

or no less,
that great old bar at 49th and First.

◊

Unbeknownst to Brian, I had also made a scrapbook. Mine was much more makeshift than his. I had taped a picture of him as a child to the cover of a cloth book I had bought at a stationery store and crammed it with letters and photo-graphs sent by friends, quotations I liked, and photographs I had found until it did not close well. I cannot remember what I wrote.

At the time, both of us were touched and somewhat disappointed, I think, in the gift that each received from the other. Each of us made the book we hoped to receive, and each noted what was absent as much as, or more than, we appreciated what was given. I wondered why the empty pages in my book were not filled. Brian wondered why I could not be more careful, why I was not organized enough to be careful.

It is only years later that I can see how the books expressed the worlds we came from, worlds that, flawed as they may have been, shaped our understanding and so our expression of love.

I grew up in a noisy, chaotic house where no one ever finished a sentence. Old copies of the *New Yorker* and *Antiquity* covered the coffee table in the den. Once when looking for a book review in an old *New Yorker,* I reached to the bottom of a pile and pulled out a copy from 1979. That was in 1997.

It was difficult to find a pen or pencil to write down a telephone number; when one was unearthed, either the ink was dry or the lead was broken. Stacks of old mail sat on the steps and on the kitchen counter, waiting to be sorted. The house was not dirty, just cluttered. For years, my mother talked about how she was going to "dig out." I have letters she wrote to me when I was at summer camp, talking about how, after digging out, she was going to sort through the slides. Thirty years later, when we talk on the phone, she tells me that this year—for sure—she is going to get to those piles of mail, and then to the slides.

Brian's parents were older than mine; his mother had been forty when he was born. A private person to begin with, she had entered a more solitary phase of her life by the time I met him, less engaged in the larger community than she had been. His parents' house was full of beautiful Oriental carpets and antique furniture. It was quiet, like a museum. Conversations were calm; sentences, always finished.

My family's piles drove Brian mad. That no one, including me, ever finished a sentence frustrated him even more. My in-laws' house, in turn, felt too orderly to me, the quiet seemed a sign of something missing, not of something achieved.

For me, love was about the feelings that went into a gift as much as the gift itself; it spilled over the edges. The book that I gave to Brian was like that. Love in Brian's household was expressed in gifts of beautiful objects, and love for Brian was in the attention to detail, in the finished product. The book that he gave to me was like that.

◊

As an anthropologist, I think about these differences. As a graduate student, I spent hours reading about economic systems and political processes, about kinship and rituals, about all the trappings of a culture that must be understood in order to establish cross-cultural communication. And yet, I never read anything about how difficult it is for two people to really see each other. If it was so hard for me to see through the eyes of this man, with whom I shared so much, how in the world did I expect to enter the world of those whose lives were completely different from my own?

5

THE BLACK BOX

September 1989

CAMBRIDGE, MASSACHUSETTS

When Lok Bahadur and I were chased from the village with the so-called indigenous conservation project, I learned that it was not enough to call land sacred or to declare that it was to be protected. What mattered was who held the rights to that land and whether others accepted that claim. More than a statement about the sanctity of the land, the lama's claim to the forest that others felt was theirs had been an attempt to manipulate the fascination of scholars and development workers, particularly those involved in conservation projects, with sacred lands. Peel back the layers of indigenous conservation practices, and there was, not surprisingly I realized (although I assumed differently at first), a struggle for power. So, after returning to Harvard, I began to focus on the seemingly mundane topic of land tenure.

I had thought that the greatest threat to indigenous beliefs and practices came from missionaries who pushed their version of God on those who had other beliefs. As I read, I began to realize that cultures had been altered just as powerfully by transformations in their rights to the land and to resources on that land, particularly under colonialism but not only under colonialism. In Nepal, which had not been colonized and from which missionaries were prohibited, this transformation had occurred more subtly, in the incorporation of rural communities—with their own systems of allocating rights to the land—into the legal and administrative framework of the central government in Kathmandu.

Like different religious views, different systems of land tenure express fundamentally distinct ways of relating to the land. In a system of common property, access to the land depends on membership in a community that, in turn, conveys the rights to graze livestock, say, and gather firewood. These rights are allocated and governed according to, in the words of E. P. Thompson, "an inherited grid of customs and controls" (1976:337). What is important is the relationship between an individual and the community, not between an individual and a plot of land.

In a system of private property, again in Thompson's words, "property must be made palpable, loosed for the market from its uses and from its social situation, made capable of being hedged and fenced, of being owned quite independently of any grid of custom or of mutuality" (1976:341). The relationship that matters is between an individual and a plot of land. Although useful, an individual's relationship to a community does not impose any constraint on his use of the resources on the land he "owns."

I chose Hedangna, the largest village in the Makalu-Barun area, as a research site because *kipat,* one of the best-known customary systems of land tenure in Nepal, was still operating there, even though it had been declared illegal several years earlier by the government. Since the land in this part of the country had not yet been surveyed, traditional systems of tenure were still in effect. To me, looking into how *kipat* worked in practice offered a more grounded and culturally appropriate path to discovering indigenous conservation practices than what had been offered by the lama of the village with the "conservation area."

Everything I learned about *kipat* before going to Hedangna came from books by the Nepalese economic historian Mahesh C. Regmi, who described it as a system of communal landownership. When I thought of communal ownership, I envisioned men, women, and children surrounded by fields and forests and open meadows. Streams and valleys, rocks and rivers marked the borders of their commonly held land. When I thought of private property, I saw solid black lines spread across a featureless landscape on a surveyor's map. Where there were people, they were usually alone: dots scattered throughout the white space on the map.

From these images came assumptions and generalizations: rights to use common property are based on an ethic of cooperation and sharing, on face-to-face interactions and a built-in sense of "enough." Private property is individualistic; it depends on lawyers and surveyors, since land is an object to be exploited and the entire system is organized around making a profit. Some of my assumptions were grounded in reality. Others were simply my bias that traditional, less-commercial ways were wiser than the practices I observed at home.

◊

A few weeks before leaving for Nepal, I drove to Yale University to speak with William Cronon, an ecological historian whose work on colonial New England had most shaped my ideas about the relationship between land and people. I had never met him. He said that he had twenty minutes to speak with me. I quickly mentioned the first of two points from his books that had most influenced my work: his comment in *Changes in the Land* that "the pig was not merely a pig but a creature bound among other things to the fence, the dandelion, and a very special kind of property" (1983:14). Like the pig, I said, the idea of conservation had arisen from a particular cultural and historical context and was bound, among other things, to industrialization, population growth, and a very special kind of property. I hoped to explore those connections, and I wanted his advice.

The second point related to a passage in *Nature's Metropolis*, about how seemingly isolated actions and events are attached to political and economic forces beyond what we can imagine. I said that this passage had made me think of the Makalu-Barun Conservation Project and the cultural and economic factors that created the need for conservation in the first place. It seemed to me that our culture romanticized wild places, even though we lived in and supported an economy that contributed to the destruction of the very places we hoped to save. By focusing on those wild places, we could avoid assuming responsibility for the ecological consequences of our ways of living at home.

I talked about how the vision for the Makalu-Barun Conservation Project had originated in the cultural and economic context of the United States. A well-funded program of an American nongovernmental organization, it had the financial and political resources as well as the moral authority needed to assert its values of preserving biological diversity and promoting sustainable economic development. Although the project was certainly about these things, intentionally or not it was also about the right to define boundaries on the land and to determine the activities that would be allowed within those bounded areas, the superimposition of an idea of the appropriate relationship between humans and the earth that had emerged largely in the Western world onto an existing relationship shaped in completely different circumstances. In many ways, conservation projects were not so different from missionary activities; since this occurred under the name of conservation, though, these projects represented a redefinition of reality that was more palatable to liberals, who would have balked at the idea of imposing Christianity on practitioners of another religion.

Cronon listened as I described my research objectives, my beliefs about the value of common property, my hopes for gathering information that would help the Makalu-Barun Conservation Project staff avoid some of the perhaps inevitable conflicts that result from doing conservation in the developing world. He nodded, and then cautioned me to look beneath the choices and decisions that people made. The New England colonists had built fences not to steal the Indians' land, he said, but to pen their pigs. The fences had consequences beyond what the settlers' ever imagined—some intended, some unintended. The point was not to blame the colonists for those consequences, but to identify the logic inherent in the colonists' system of property, a logic that could not coexist with the Indians' ways of allocating rights to the land. The deeper misunderstandings, the blinders that prevented each side from seeing what was on the other side of the fence, were what were important. The most I could hope, he said, was that by documenting the villagers' ways of conceiving of and relating to the land, I could make the Makalu-Barun project staff hesitate before implementing programs that they would almost inevitably carry out.

Just before my twenty minutes were up, he told me that we all approach our work with analytic frameworks, with a black box that allows us to interpret the world. Things that fit inside the box make sense; those that do not, we ignore. When I encountered material in Nepal that did not fit into my box, he advised, discard the box, not the information. He urged me to let myself be transformed by what I discovered, not just defend what I already knew. I nodded, assuming that what he said would be easy.

6

THE BARUN FESTIVAL

January 1992

Brian was wearing long underwear when he got off the Twin Otter at the grass airstrip in Tumlingtar. It was January, and he had come from New England, where long underwear made sense. It was 60 degrees in Tumlingtar. I had not told him what to expect, he said after we hugged, briefly, surrounded by villagers and porters watching to see how we would greet each other. He had had to guess as best he could.

We had not seen each other for four months. We had had no contact since I left Kathmandu for Hedangna a month earlier. Mail was undependable, and so we rarely wrote letters. In 1985, when I had lived in Nepal for eighteen months, Brian and I wrote ten-page letters to each other every week. He told me about his work and his dreams; mostly, he expressed support and encouragement for what I was doing. I shared with him my experiences, my reflections, and my loneliness. These letters kept our relationship alive and real. They allowed each of us to open to the other in a way we had not before, bringing us closer even though geographically we were farther apart than we had ever been. Now, in 1991, no communication had had the opposite effect.

Brian took off his long underwear, and we began the three-hour hike up the dusty trail to the Khandbari bazaar, where we would stay for the night. As we walked, we talked about what we had seen and done during the past four months. I tried to describe Hedangna. Brian brought me up to date on news about his work and our home, about my family and his. Despite the distance between us, the sense of familiarity and comfort and the ease of speaking a language without thinking allowed me to begin to relax. I felt lighter and more carefree than I had had in months.

It was dusk as we approached Khandbari. The trail widened as we came to the edge of the market town. Single-story wooden huts with thatched roofs were replaced by two-story wooden buildings with tin roofs. We passed shops where Brahmans and Chetris sat on low benches, drinking milk tea from glasses and smoking factory-made cigarettes. Single bulbs shone dimly overhead. Women bent over cooking fires, heating water for tea and boiling lentils for the evening meal. We continued to walk until we reached the center of the bazaar and the Arati Hotel, where I had a room.

The Arati Hotel was one of two hotels in Khandbari and the one where foreigners usually stayed. Cement steps led up from the dusty path into an open room painted turquoise blue and filled with blue tables and chairs, a blue that made me feel as though I were sitting at the bottom of a cheap, empty swimming pool as I ate. I chatted a bit with Nabine, the owner, who sat at what seemed to be a speaker's podium and stared past the restaurant and into the square beyond. Brian then followed me through the blue restaurant; down a dark, grungy hallway; up a very steep outdoor staircase; down another narrow hallway, a bare bulb showing us the way and the linoleum peeling in the corners; and into my room, also painted blue.

For me, moving from solitude to relationship was difficult. It was not always easy being on my own, but once I got used to it, it seemed easier to keep doing things myself than take the time and make the effort to include someone else. Transitions never seemed difficult for Brian. He was always immediately present, eager to see me, curious about what I had been doing, and ready to join in as well. Each time we reunited, he came bearing gifts: cherries and champagne, grapes and M&M's. He enlarged and framed photographs for me to hang on my walls. Worrying that I would bring the wrong gift or the wrong food, I often arrived empty-handed, missing the point of the gift. I was confused by the difficulty of this simple gesture; Brian was confused by my distance.

It was my birthday. Brian unpacked smoked salmon and Kahlua to celebrate. We took off our packs and lay on the bed in the turquoise room, eating the salmon and drinking the Kahlua, talking late into the night until the dusty square was quiet and the houses were dark, trying to find our way across all the months we had been apart.

◊

I had planned to spend a day looking around Khandbari, so Brian could rest before a long day of hiking. But the next night was the Barun *mela*, an all-night festival held at the confluence of the Arun and Barun rivers, one of the biggest festivals of the year. These two rivers are believed by Hindus and Buddhists alike to be sacred, and villagers travel from throughout the region to make offerings to them on designated days each year. The goddess of the Barun accepts milk in a vessel of cow manure at dawn the morning after the festival. I had hoped to attend both the festival and the ritual. Because Brian's plane had been delayed, we had only a day and a half to hike a distance that would take trekkers, could they have traveled on this route, four days and Nepali civil servants three.

I described the festival to Brian as we sat in our blue room above the town square. I said that it would be a long, hard walk to get there in time, but did not leave much room for him to suggest that we not go. Brian replied that it sounded wonderful and worth the effort. We finally crawled into bed and, just before dawn, a few hours after we had fallen asleep, awoke, repacked our packs, and walked into the dark empty square and up the dirt path through the slowly wakening market town toward Hedangna.

◊

All I remember of that hike is the last hour in the dark, trying to make our way along the steep rocky trail by headlamp. It was one thing to have agreed to attempt this trek the night before, in our turquoise room with stomachs full of smoked salmon and Kahlua. It was quite another to hike all day on a rough, rocky trail with no time to rest. Brian wanted to stop for the night at a tea shop we had passed at dusk. I knew that I was pushing too much on the first day. And yet we also had to take into account the length of time it took to cook food on an open fire. If we did not reach Num that night, we would not make it to Hedangna the next morning in time to find Jaisita at home and rice that had been cooked. We would never reach the Barun River the next evening if we had nothing to eat all day. This was the one chance we had to see the Barun festival, and it had seemed worth doing all we could to make it. But now, stumbling along the trail in the dark, trying to convince Brian, I was not so sure.

At home, Brian was more decisive and far more assertive than I. Making decisions was painful and difficult for me. The choices seemed endless, as did the possibilities for making bad decisions. I could never trust that I would not change my mind. Because Brian seemed so clear about what he wanted, it was simpler to let him decide. But in Nepal, there were fewer options. I was also clearer about what I did and did not want. I felt confident in my ability to get the things I needed, and I was more assertive. Brian was not on familiar territory, nor did he speak the language. Although he continued to question whether I could secure the most comfortable sleeping arrangements, the best food, or the most *jad*, he began to look to *me* to make decisions as often as he relied on himself. And so, surprisingly, did I.

I suggested that we stop to eat a Snickers bar. As we took turns taking bites, I again tried to describe what I had heard about the festival, tried to explain why it seemed so important. In his mind's eye, Brian had only the images I gave him and, in his stomach, only the white rice and watery lentils eaten hours earlier. What he did know was that we had been walking since dawn; we could hardly see the trail; and I kept promising a village that still had not appeared, something I had been guilty of in the past. He again asked how much longer it would take to reach the promised village and said more insistently that he thought we should turn back to the tea shop we had passed some time earlier.

For me, it made no sense to turn back. We would never get to the festival on time, and then the whole day would have been a waste. I pulled out our last Snickers bar. Brian conceded, and we set out toward Num.

We reached the village forty-five minutes later. Num was a collection of rundown tea shops that catered to civil servants, heading north and south on the trading route to Tibet, and to trekkers, much fewer in this part of the country than in the Everest and Annapurna regions, heading northwest to Makalu base camp. We ate ramen noodles, drank sugar tea, and then slept restlessly on narrow wooden beds, again rising before dawn to set out for Hedangna.

The trail that heads north from Num drops steeply to the Arun River, about a two-hour descent, and then climbs more gradually on the other side, where Hedangna is spread out on an east-facing ridge. We arrived at Ganesh and Jaisita's house on the far side of the village just as Jaisita was heading to the fields. The coals had gone out, but anticipating our arrival, she had cooked enough rice and spinach for us to eat. She came back inside to serve us food and *jad*, watching everything Brian did and hardly saying a word.

We ate quickly; repacked our bags, taking only what we needed for the night; and again started to walk. As we reached the ridge just north of Hedangna, we joined a line of villagers walking north up the Arun Valley: women dressed in brightly colored lungis, not yet faded from use; men in Nepali tunics and pants saved for special occasions; and teenage girls in groups, giggling as they walked. Everyone carried a woven shoulder bag that held a bottle of *raksi* and some uncooked rice for the evening meal. They greeted me and stared at Brian, commenting on his shoes, his pack, and his mustache; they asked who he was, and then continued on their way.

We bumped into Devimaya's mother, Dhanmaya, gripping the hand of Rendha, her seven-year-old son, and practically dragging him up the trail. She was leading him back to the river that she and her husband had visited ten years earlier, to pray for a son, after their two sons had died. She looked regal: a bright purple lungi with a golden dragon circled her hips and legs, she wore a velvet maroon shirt, and a white shawl covered her dark hair. She was bringing Rendha to the festival to express gratitude, so the goddess of the river did not think that Dhanmaya took her gift for granted or would not notice if she chose to take him away. Looking serious and small in his newly sewn, robin's egg–blue Nepali pants and shirt and olive green–checkered jacket, Rendha stumbled to keep up. He clutched a slingshot bought for 15 rupees the day before at another bazaar, probably the first time in his life that he had received such a gift.

"*Barun ta leye?*" Dhanmaya called out to me in Yakabha. "Are you going to the Barun?" "*Barun ma lenge!*" I replied, happy to see people I recognized, happy because I knew that we would make it, happy because Brian seemed happy as well.

◊

Just north of Hedangna, the Arun Valley narrows and the trail drops close to the river— at times climbing steeply up or down through the thick forest, at other times passing through open terraces farmed by villagers whose homes are farther up the ridge. Except for the village of Gola, a lonely police checkpoint with two or three run-down wooden shacks, we did not pass any permanent structures, only grazing shacks built at the edge of fields.

Beyond Gola, the valley becomes even narrower and the trail winds steeply through the forest, squeezed between the river and the ridge rising directly from the valley floor, blocking the sun except for a few hours each day. Just as the sun was setting, we climbed down the last hill and saw a line of leaf-covered shacks built from freshly cut saplings earlier that day and spread across several wide terraces along the banks of the river. Mist, rising as the temperature dropped, mingled with smoke from fires over which goat meat fried and water boiled for tea.

We entered the first leaf-covered hut and were immediately overcome by the smells, the noise, and the crowds. Kerosene lamps shone on food for sale: biscuits, candies for 1 rupee, and *raksi*. Lots and lots of *raksi*. Men gathered in groups to gamble with cards and dice. More than thirty years earlier, these games were banned by the Rana prime ministers except on certain holidays. Children mimicked their elders, using sticks and leaves instead of cards and dice. Voices in Nepali, Tibetan, and Rai shouted over the steady roar of the Arun and Barun rivers, which come together near the path and made the bustle between the shacks feel more chaotic and crowded than it was. Brian and I did not speak but moved with the crowd, in awe of the smells and sounds, watching, watching, walking, walking. We wandered until we bumped into Raj Kumar. He showed us where to drop our packs, and then joined us in a glass of *raksi*. We then wandered some more until we met someone else I vaguely knew, whom we joined for another glass of *raksi* and a plate of fried goat meat.

Outside the string of leaf huts, on the terraces closer to the river, Bhotiyas from the north gathered in large circles and began a rhythmic chant and dance under the stars. We walked over and stood silently at the perimeter of the circles to watch them, transfixed by the chanting, a sound that seemed to contain the expanse of the Tibetan Plateau, the smell of burning juniper, the darkness, and the roar of the river at our backs. Brian reached for my hand.

◊

I thought of the night of the full moon at Bodh Gaya in northern India, where Brian and I had gone to see the Dalai Lama bestow the Kalachakra Initiation on a crowd of almost 300,000 people. As the hot sun dipped behind the hazy plain, pilgrims flocked to the site of the *bodhi* tree under which the Buddha is said to have attained enlightenment. A huge stone temple built to commemorate the site towered against the dark sky. The moon shone on throngs of people walking around the base and chanting *"Om Mani Padme Hum."* Young monks, old women, and teenage boys and girls placed flickering candles on the stone walls, stuck incense into the cracks, and poured melted butter into the thousands of butter lamps placed at the base of the stupa. Others prostrated themselves on wooden platforms, chanting, their hands and knees wrapped in padded cloth as protection. Large groups of monks sat at the base of the stupa, also chanting. The air was heavy with the smell of burning butter, and the entire compound was alive with the physical manifestation of so much faith. The chanting, the candles, the offerings of food piled in front of the stupa—all of it arising from faith.

Now, at the confluence of these rivers that were gods, we once again encountered the magic and mystery of something unknown—of something that could not be known. There was nothing to do but pay attention. The distance I had felt between the villagers and me faded. The differences between Brian and me fell away. We were close in the way we were close in wild places, in awe of the landscape, natural and cultural, a sense of wonder that spilled into how we felt for each other. There was a timeless quality to the closeness that, I believed, made up for the time apart, imagining that a response to an opening in the landscape was the same as creating that opening ourselves.

◊

Most people stayed up long into the night, eventually lying down wherever they found space, some with a blanket, some without. It was January in the mountains, and although it did not snow at this low elevation, it did get quite cold, especially at night. Brian and I spread our sleeping bags on the ground in an empty field away from the center of the festival, and dozed off to the hum of voices and singing and the constant roar of the rivers.

We woke in the gray morning light, long before the sun climbed over the mountains on either side of the narrow valley. It was cold and damp. We packed our bags and followed the men, women, and children who were walking toward the boulders on the bank of the Barun to prepare their offerings. The Arun—the wider, older, and stronger of the two rivers—cuts through one of the world's highest mountain ranges and is believed to bring health and good fortune. Villagers hold feasts in their own villages to honor the Arun.

The milky green Barun—narrower, younger, and more turbulent— tumbles through a chasm and circles slowly in a pool before merging with the Arun. The spirit of the Barun is said to be female and, unlike the spirit of the Arun, does not eat meat. Instead, she likes offerings of milk, preferably served in cups made of cow manure. She also accepts flowers placed in boats made of leaves, but this gift is offered at her source on a pilgrimage undertaken in August, at the time of the full moon.

Only the hardiest villagers could make the trek to the headwaters of the Barun. Many more people attended the winter festival to make offerings at the confluence of the Barun and the Arun. The night before, everyone had interrupted others with no concern about intruding. This morning was different. People nodded in greeting but did not speak, or when they had to, they whispered. Men who had been loud and drunk the night before were silent and serious as they helped their wives press a clump of cow manure into the shape of a cup and fill it with milk carried in a wooden container from home. After placing the offering on a rock by the edge of the river, the couples knelt side by side, closed their eyes, and pressed their hands together in prayer. Farther downriver, a *baba* (Hindu holy man) sat cross-legged on a narrow flat beach, wearing a bit of orange cloth, his long white hair in dreadlocks, lost in his chanting. Offerings complete, men and women returned to the now disheveled leaf-covered huts, gathered their belongings, and began the four- or five-hour walk home.

Nepali festivals, the Barun *mela* included, are not completely magical. Everyone drinks too much, especially the men. And they fight, usually with fists but sometimes with *kukuris,* long curved knives made famous by the Gurka soldiers and still used for everything from cutting fodder to decapitating goats. Once Brian and I left another festival at dawn, shaken and frightened by a drunken man careening through the center of the clearing, waving a *kukuri* through the air. The day after the Barun *mela,* at another festival held on the banks of the Arun just below Hedangna, a man was almost killed in a fight with someone wielding a *kukuri.* And five years later, Dev Kumar, Devimaya's fiancé, would fall to his death while walking home, drunk, from the Barun *mela.*

I did not know any of this the night Brian and I wandered through this festival of flickering lights, shadows, and chanting in a place that was usually dark and quiet. We were close to something neither of us tried to put into words. When I try to remember that night, what I saw and how I felt, I mostly remember the glow, not just of the kerosene lanterns, but of the feeling that comes when the veil between worlds is lifted, when it seems easy to step across the boundary and enter the world beyond.

◊

At some point in our lives, so the sacred traditions say, we receive a call to leave the known world of our family, our community, our culture, led by an intuition of our place in the world, a "beckoning uncertainty" as poet David Whyte (1991) says, that draws us deeper into our destiny. To respond to this call—which we can choose not to do, but rarely consciously choose to do—requires risking everything we hold dear, although we may not realize it at the time.

The Barun *mela* was a gateway into another world, a world where the things that mattered were not necessarily the things I had been told were important. I imagined that I might be transformed by responding to this beckoning uncertainty, but I assumed that the box I climbed out of would be able to hold me when I was ready to return.

7

THE BAMBOO BRIDGE

January 1992

The only thing I ever "cooked" on my own in the eighteen months I lived in Hedangna was water for tea. I boiled the water on my camp stove in a corner of my room. While the black Nepali tea leaves steeped, I dipped the corner of my towel into the leftover water and then pressed the towel to my face, drawing in every last bit of heat, imagining my whole body steeping in the moist warmth.

More than a year later, I was on an airplane flying home, my research complete. The flight attendant passed out a steaming white washcloth, and, for a moment, washcloth pressed against my face, I was back in Hedangna. It was early morning. I was dressed in a lungi, long-underwear bottoms, and a worn gray Synchilla jacket. I had a woolen hat on my head and a cup of hot tea at my side.

◊

After the Barun festival, Brian and I returned to Hedangna, where he planned to stay for the month. We quickly settled into routines that were different but similar. Brian woke at dawn to take photographs in the village. I did yoga and then copied my notes of what I had observed or conversations I had had the previous day before heading out to spend time talking with villagers before they left for their fields and forests for the day. After returning for a mid-morning meal of rice and cooked greens, we again wandered through the village. I stopped to talk with anyone I could find who was willing to talk, while Brian took photographs or just observed, happy not to have his days planned. When there was work that both of us could do, which was not often at this time of year, we joined people in the fields to help. Some days, we went to the stream to wash clothes and bathe. Our lives had a quiet rhythm, with a slowness and sameness that felt healing. It created an intimacy that was different from that experienced during our trip to the Barun festival, but similar in effect.

◊

A few weeks after arriving in Hedangna, I tried to conduct a very basic household survey, which my advisers had recommended I do early in my stay in the village. I did it less to collect accurate information than to begin connecting faces and names with houses. I quickly found that instead of bridging the distance between the villagers and me, the survey simply made that gap more explicit. I would stand in a mud courtyard while the householders sat cross-legged on a mat on their porch, staring at me and talking among themselves about my clothes, my bag, my shawl, my shoes, my hair. They would ask whether I was married and, if so, where was my husband and why didn't I have any children. They asked whether I was connected to the Spanish school rumored to be coming to Seduwa, a village three hours to the west, or to the national park, and was it true that the park had been created so the government could capture wild animals to ship by helicopter to Kathmandu, where they could be sold to foreign governments for lots of money.

I answered their questions, and then asked if I could ask them one or two as well, questions regarding their names and family relations. They wondered why I had to know their names and what I was going to do with the information I wrote down: Was I taking it back to my country, and, if so, what would I do with it there? When I tried to change my questions to a subject less threatening than asking their names seemed to be, they told me that they did not know the answers, that I should go to their neighbors' houses and ask them, and that in any case they did not have time to talk. At that, they would pick up their tools and head down the hill.

Each morning before setting out to a different section of the village, I would give myself a pep talk while I drank my tea, trying to convince myself that today people would talk to me, that a survey really was the best thing I could be doing, and that the information I was gathering was, in fact, useful. But finally I gave up and let myself do what I had wanted to do from the start, join the villagers in the fields and forests to help in whatever way I was able. No one ever turned down my offers to help.

I had arrived in Hedangna in December, at the end of the rice harvest, when there was plenty of work to be done, even by unskilled hands. Brian arrived in January, the coldest and rainiest month with the least amount of work and few demands for unskilled labor. Men and, especially, women spent hours in the forest splitting wood for the year's supply, and then for the next few weeks women carried the wood home in baskets on their backs. I kept asking if I could help, but Ganesh, Jaisita, and Raj Kumar's brother Deuman told me that the work was too hard, that I would not be able to do it. Finally, after an afternoon spent writing in my room and watching Jaisita carry load after load of wood into the courtyard, I decided that there must be some way I could help with the wood, if only by carrying a few loads. So the next morning, when Brian went to the school to spend the day with Raj Kumar, I went to the forest with Jaisita to carry firewood.

◊

Jaisita, her father's wife, and I left soon after eating, an empty bamboo basket attached to a tumpline around each of our foreheads. Their twelve-year-old daughters joined us, also carrying bamboo baskets. We walked slowly up the trail, so slowly that I thought we would never reach our destination before nightfall. We had not gone very far before we stopped to rest on a stone platform. Jaisita and her stepmother slowly rolled cigarettes. As they inhaled in short, tight sucks, they talked about who was doing what in the village. After some time we started again, but soon we came to another stone platform, where we again stopped and they again rolled cigarettes. This break was even longer. After they finished smoking, each pulled her daughter over and roughly picked lice from her hair. Then Jaisita took off her own shirt and sat bare-breasted while she picked lice from her armpits and from the seams of her shirt, where they liked to gather. I was bored and impatient, thinking of what I could be doing had I stayed in the village. During this second forty-five-minute break, I decided that all of Ganesh and Jaisita's talk of the *dukka* (hardship) of life in Hedangna must be an exaggeration.

We eventually left the main trail and climbed up a steep, narrow, and rocky path through scrubby land overgrown with ferns and *bon mara,* an invasive plant that was taking over uncultivated land, and eventually into the forest that grew close to the top of the ridge north of Hedangna. We did not stop until we reached a small, steep clearing by a tree that had been cut down. Jaisita, her stepmother, and I began to collect dead wood from the forest floor while the girls were sent to gather fodder for their pigs. After filling our baskets, we returned to the tree. Jaisita looked through the pig food her daughter had collected, sent her off to get some more, and then picked up an ax hidden by the trunk of the tree and began to split wood.

I got my load ready to carry back down the hill, and when Jaisita left to gather another load of dead wood for her daughter to carry home, I picked up the ax and tried to chop some wood, hoping that it might be a way to help. It was unlike any wood I had ever split, and I had split plenty of wood. I could not even make a dent in the top. I ended up shredding the wood instead of splitting it. The rough handle of the ax tore the skin on my hands. I heard voices coming closer and quickly stopped before anyone could see me. When they returned, I hid my hands so Jaisita could not see that they were bleeding. Jaisita then said that she was going to stay to split wood, that there was not anything else for me to do, and that I should return to the village with the girls.

I was petrified at the thought of walking down the steep, narrow, and rocky path with a load of firewood. But I had insisted on coming, and I was determined to follow through. I watched the young girls lift up their baskets and tried to do as they did. I braced my basket against a small tree, sat down—bottom on the ground, knees up, basket against my back—and placed the aloe tumpline on my forehead. I brought my forehead down and forward to avoid straining my neck, and carefully and shakily stood up. I adjusted the basket on my back and, clutching the straps on either side of my neck to steady the load, slowly and carefully took a step. I could not do it.

I sat back down to take some wood from the basket and lighten the load. Watching as the girls scurried down the trail with loads almost as big as the one on my back and disappeared into the woods, I again arranged the tumpline across my forehead, basket against back, and tried to stand up. This time, the basket flipped over, spilling wood across the ground. Jaisita, who was watching from up the hill, came down to help me. It turned out that I had the basket backward against my back, and the strap was twisted. She showed me how to fix it. I shakily stood up, took a step, slipped, and sat down on my bottom. Jaisita, who had already headed back up the trail, saw me and burst out laughing.

The third time, it was easier to reset the basket against a rock, sit on the ground, place the tumpline on my forehead, and stand up. I managed to stay standing. With relief, I heard Jaisita call to her daughter to wait for me. The last thing I needed was to be lost in the forest with this impossible load of wood. Trying to keep my head and neck steady, so the load would not swing around and make me fall, and taking baby steps to keep from tripping, I followed the two slight girls, their loads of wood more than half their size, scampering down the trail. It took all my concentration to make my way through the rocks and roots, and I clutched the aloe straps with both hands to relieve the pressure on my neck. The girls, though, were in the midst of a brisk business of buying and selling grass bracelets, pausing every so often to stoop and pick leaves off the ground to use as money and chattering the whole time. They stopped every few minutes to wait for me to catch up, and then ran off down the trail.

When I had confidently said that I could help carry firewood, I had imagined that it would be like carrying a heavy backpack, which is difficult and exhausting but never really dangerous. The baskets filled with firewood were awkward and unbalanced, a slight turn could make them spill over, and the pressure on my head and neck was unlike anything I had experienced. Muscles or tendons I never knew existed ached for a week after I carried that load of wood. Later when I told Ganesh that the trail was hard, he nodded with a seriousness and directness he rarely expressed, and said that someone who slipped while carrying a heavy load like that could die.

That day carrying firewood helped me understand what Jaisita had meant when she shrugged in response to my question about what was good about this place. Firewood is far away, she would say. Water and fodder are far away. Everything we need we have to carry on our backs. What is good about it? A trekker once commented on the beauty of the hills to a friend who worked in western Nepal. Not if you have to walk across those hills to get home, my friend replied.

When I got back to my room, I found Raj Kumar talking to Brian. Raj, like Ganesh, simply nodded knowingly when I rubbed my neck and talked about how difficult my day in the forest had been. I should not do that sort of *dukha*, Raj told me once again. Then, as he left, he reminded me that Chute Rai, his father's older brother, was always at home during the day. Chute had said that he would be happy to talk to me and that he knew a lot about the past. It was easier than carrying wood, Raj added, and headed home.

◊

Chute Rai was sitting on a straw mat in his small courtyard, slicing thin strips of bamboo to weave into a basket, when I walked up the next morning. He wore a torn green woolen sweater, cotton Nepali pants, and a dirty white woolen hat. His face was dark and wrinkled, his dark eyes disappeared into thin lines when he laughed, and he had a thin gray mustache. His hands were gnarled, and he had a loud, hacking cough.

He glanced up as I approached and asked if I was walking on or coming to see him. I replied that I had come to see him. He nodded, stood up stiffly, and walked to the porch to get a straw mat for me to sit on. He sat back down and asked when my husband was going home. As soon as I had answered, he said that I was here to write a book, wasn't I, and that I had to study hard. I nodded, and then he asked if my country was like Italy, or if it was better. He had been to Italy, he told me, not waiting for my answer, to fight in World War II. He had been training in the Indian army when it was still under British control and had been sent to Italy. He had taken a boat from Bombay to Iran and had traveled by train the rest of the way.

As Chute spoke, I tried to imagine what it must have been like to leave Hedangna as an eighteen-year-old, never having seen a road or an automobile or even a three-story building—and suddenly finding myself not simply in an entirely different world, but in an entirely different world that was at war. I asked what it had been like for him, what he remembered. The two things he mentioned, the same things he mentioned every time I saw him, were the long pieces of bread—he would stretch out his arms to make sure I understood exactly how long—and the smooth, flat roads.

I asked which he liked more, Italy or Hedangna. Without pausing, he said, "*Hernos, bahini,* look, younger sister, the *bikas* [development], and food and things of Italy were very good, but your own village is always better. Even if you don't have any money in your own village, you have relatives, older and younger brothers. The food in your own village is always better," he said. "The water is better in the place where you are born."

I asked what it had been like to return to this village after seeing and experiencing all that he had experienced. He shook his head and said that he was a fool for having come back when he did, that he should have stayed in the Indian army to earn money and to travel. But his mother had wanted him to come home, he said. It had not seemed that he was making much money at the time, and so he had returned to grow rice on his father's land.

The weather was good in Hedangna, he said, not too hot, not too cold; the village was not too high or too low. But the villagers had to plant crops just to get food; they had no money, no way to earn money, and no way to save money. "*Hernos, bahini,*" he said again, "you have to do business now, while you can." He asked how much money Brian earned a month. I told him a figure that was about half of what he actually made. Chute nodded and said that it was a lot, so I explained how much it cost each month to live. He nodded again and said that we had to save money each month. He repeated how much I had said we spent each month, calculated how much would be left over from Brian's salary, and told me to put that money in the bank. I felt slightly guilty that he was doing all this work with false numbers. He said that we had to save that much each month until we were fifty, which, he figured, was as long as we could work. Then we would have that money to care for ourselves. He then began to calculate, based on how old I was and how many years of work we had left, how much money that would be.

He then sighed and said, "We can't do that here."

"But your sons can look after you," I said.

"No, they can't," he replied. "They have to raise their own children. I am an old man. *Hernos,* please look."

Chute began to cough again. I wished, not for the last time, that I was studying to be a doctor of medicine instead of a doctor of anthropology, wished that I could offer him a cure in exchange for all he offered me.

Chute asked whether my parents were still alive and how old they were. He nodded and said that my father was his age, and he asked what he had done for work. He told me that I had to help them, that it was good dharma for me, and he asked if Brian had parents, how much land they had, and if he had brothers. He asked if I had brothers, and when I answered that I had one younger brother, he asked about his education level and the kind of work he did. He did not ask about sisters, who were not relevant to the topic at hand. My brother would care for my parents when they were old, and Brian and his brothers would compete for property left by their parents, which would, in turn, affect our livelihood.

He then asked what information I needed to know. I told him, and he proceeded to list the names of everyone in his clan, from the first ancestor seven generations ago, by heart. He spoke slowly, making sure I wrote down the names correctly. He told me that I had to learn the language well so I would understand everything, and then had to visit each house and get the information I needed. If I was going to write a book, he repeated, I had to make sure I got all my information right. As I stood up to leave, he told me to come back again soon.

◊

Chute had a cough most of the time I was in Hedangna. Like most older villagers, he did less strenuous but equally essential work: watched goats, kept chickens from pecking the rice that was drying on woven mats, dug potatoes, wove bamboo baskets, and cooked rice so it would be ready when his wife returned from the fields or forest. I visited him several times each week; we would chat a bit about the weather and what Ganesh and Jaisita were doing. He would ask how much money Brian made and again tell me how much we should be saving in the bank. He would say something about Italy and bread. Then I would ask him to clarify or confirm what I had just learned about land surveys or taxes or the oral tradition.

Chute was the first villager other than Raj Kumar who understood something of what I was doing, and the only villager who called me over during a ritual to ensure that I saw the things he thought were important for me to see. He always checked to make sure I was taking notes of what he said, instead of suspiciously asking what I was writing and why. When gossip came up, he would grin, his eyes closing to a slit, and tell me to ask the person involved, so that he or she could explain for him- or herself what had happened.

◊

After speaking with Chute, I would find Brian and we would go for a walk or sit on a stone resting platform and exchange stories about our day. The first month that Brian spent in Hedangna was one of the most relaxed times we ever spent together. Brian worked very hard at home, work that pulled him into a world that interested me but that also took a tremendous amount of his time and energy. We shared an interest in understanding the connections between land and people and a desire to do what we could to strengthen that connection, but we did not share much time. In Hedangna, the one thing we had plenty of was time. His concentration on photography drew him more deeply into life in the village; it provided him with his own relationships and experiences in the village and so simply created more for us to share.

Brian is very good at finding the words or images that communicate a larger story in a way that moves people. He is good at making a point and is a compelling and charismatic speaker. I am drawn to details, to the endless variations in experience and perceptions, not because I find details inherently interesting but because I am intrigued by the way every thing connects to every other thing. My adviser at Harvard once told me I made the mistake of believing that everything that happened in Hedangna mattered. "It doesn't," she said. It was relevant only to the extent that it shed light on something else. I nodded at the time, like a good student, even as I disagreed inside, sensing without being able to articulate that the meaning I was looking for could be expressed only in the details—in the details themselves, not because they made a larger point. And that was the point I was seeking.

If I had photographed the village, I would have taken a series of images that could be lined up to capture the entire landscape, afraid that if I selected one or two photos to represent the whole, I would have missed something, capturing only the things I assumed were important and not the things that really were. If Brian had conducted interviews in the village, he would likely have stopped when he had one that told a good story, the story that illustrated the points he wished to make.

Our differences complemented each other. Brian read what I had written or listened to what I had learned and asked questions that helped me clarify the point I was trying to make. He drew parallels and gave examples from his own experience in land conservation in the United States. This helped me think about how the details I was discovering did and did not connect with the larger whole and forced me to try to articulate the meaning I had often failed to express. I liked to think that my understanding of the reality and complexity of the villagers' lives helped Brian see that rural life was not as idyllic as he wanted it to be, that there were more perspectives and dimensions to farming than he had considered from afar.

While Brian was in Hedangna, we began to talk about writing a book together, a kind of coffee-table book, Brian described, but more than a coffee-table book, one in which my words told a story that paralleled the one expressed in Brian's photographs. We would focus on the Makalu-Barun Conservation Project, gathering stories and images from the entire region, not just Hedangna.

Intellectually, the idea made great sense to me. I was interested in writing for an audience beyond academics but was unsure how to proceed. From Brian's perspective, there were plenty of photography books about Nepal, and my research would give his images a different twist. It seemed like a perfect way to combine our talents and interests to create something that neither of us could achieve alone, and it felt good to have something solid to hold us together during all the time we would be apart.

The project focused our time together. As we wandered through the village, we came up with ideas about groups of people for me to interview: porters who worked on the trekking route to Makalu, staff of the Makalu-Barun Conservation Project, lamas in Tibetan villages to the north. We began to put together a book proposal and decided that when Brian returned to Nepal the following September, we would hike up the Arun Valley to the Tibetan border to gather information.

Despite my initial interest in the project, I never brought the enthusiasm to it that Brian hoped I would. While on the trekking route to Makalu, I interviewed porters long enough to realize that I was not drawn to their stories the way I was to those of the men and women I was coming to know in Hednanga. Brian and I talked about the idea only when he brought it up, which frustrated him and confused me. He felt that I was not trying hard enough, was not willing to invest energy in a project that clearly would be good for us, individually and together.

Looking back, I see that the book we envisioned about the Makalu-Barun region was really two books. Mine did not need photographs. Brian's did not need words, at least not those that I was likely to write. Like the books we had made for each other before my departure for Nepal, this difference masked a deeper difference. Each of us had a vision about the best relationship between people and the earth; because we used the same words to describe our visions—"people" and "land," "nature" and "culture"—we assumed that they were the same. By focusing on the overall vision, not on its particulars, we could believe that we were in agreement.

I now see that this difference marked two approaches to stories. I was drawn to novels and books in which events unfolded in ways I could not anticipate at the outset, which offered the unexpected and a sense of discovery. As an anthropologist, I was bound by facts in ways that novelists are not, but I could still follow the different threads that came along, leading me down paths I had not expected to discoveries I had not foreseen.

Brian's work was concrete: he bought land and protected it. He had certain goals he set out to achieve, so he was looking for stories that supported and helped illustrate why those goals were important. In many ways, his approach was the antithesis of what I had come to Hedangna to do.

Yet Brian's vision was compelling; he articulated it clearly and convincingly. I thought that there would be nothing to lose by doing my part to make it real, and I cared more about maintaining my relationship with Brian than about the ideas or feelings I had about culture and the land. I did not realize that the tension between us ultimately touched on our openness with each other and receptivity to what was around us, our ability to be present to what was so, not to what our stories told us was so. And I did not grasp that this difference was one that mattered.

◊

One day in Hedangna, Brian and I made plans to accompany Devimaya to Dhanmaya's mother's house in a tiny village on the ridge across the Arun River, where we would spend the night. The land to the east of Hedangna is more rugged and wild than that around Hedangna, and I was curious to see it. And the trip would be a nice change from our routine. The most direct path crossed a bamboo bridge like those once common throughout Nepal: long, narrow bamboo trunks slashed together, with low railings made from rope. People said that the bridge was scary, but that focusing on something on the far side of the river, rather than looking down at the white water rushing below, made crossing it easier. We decided to give it a try.

This was another plan that sounded fine when we were sitting around a fire drinking *jad* and another thing altogether when we were standing on the banks of the river and trying to muster the courage to take the first step. Brian took one look and said that he was not going to cross it. "It is too dangerous," he said. "It isn't worth it."

Devimaya said that she would carry our knapsacks and nimbly stepped onto the bamboo trunks, bending down to hold the railings as they lowered toward the center of the bridge. The whole bridge shook as she continued across. I was terrified, but I thought that if I just followed the villagers' instructions, I was likely to be okay. Brian refused to try it. We argued about what to do, Devimaya turning away as we raised our voices. Eventually, I gave in, probably with relief. Devimaya, who had no reason to cross now that we were not going, headed back to her house up the hill while Brian and I sat by the bridge, arguing.

Years later, I have no idea what we fought about or why it became such a big deal. I wonder why I did not agree that Brian should stay behind while I went ahead with Devimaya. I doubt that he would have minded. I had no right to force him to cross a bridge that he did not want to cross. There was also no reason for me not to go simply because he did not want to.

As with the book project, I believed that our relationship depended on our doing things together. I was forcing it to fit my own expectations and so missing out on what it was, sticking with my version of the story rather than following where the characters in that story wanted it to go. Only much later did I realize that by not taking the time to clarify and assert my own perspective on these projects, I gave up a part of myself that our relationship depended on to stay healthy. I achieved an external unity but only by deepening an internal division. By letting or making Brian be responsible for the larger vision—which I was too frightened or lazy to come up with on my own—I counted on him to move in ways I thought I was unable to move on my own, trusting him to know what I assumed I did not know myself.

PART TWO
initiation

8

STORIES AS BOUNDARIES

March 1992

Brian returned to the United States at the end of February. After he left, I turned my attention more fully to my research. Although seasons continued to structure the rhythm of my days, some part of which I spent helping in the fields or hauling firewood and water, my thoughts became shaped more by my concepts about research than by types of work. In organizing my ideas about what I had to know to understand the villagers' relationship to the land, I made a diagram of three overlapping circles that I labeled sociocultural, political, and economic. Information on land use, agriculture, and trade, material that fell into the economic category, was the most straightforward and, for me, the least absorbing. The political circle included data on land tenure and taxation, which was to be the focus of my research. I was hesitant to begin by asking about politics and property, though, fearful that my questions would create as much suspicion as my survey had done. I was most interested in the sociocultural category, which to me meant exploring what the land meant to the villagers, how they conceived of and related to the sacred, and how these beliefs affected their actions and their lives. And so when the time came to begin formal interviews, I decided to start by talking to the shamans and priests, the healers in the village.

One morning I approached Kharka Bahadur, a thin, distracted priest who was my neighbor, and asked him to explain the ritual he had recently performed to call for rain and a good harvest. He agreed to meet me that evening at the house of Ganesh's cousin Amrit, where he was to perform a short ceremony. When I arrived, Kharka was standing in the dark smoky room, tossing bits of dehusked rice into the air above Amrit's head, and mumbling a chant. I was invited inside, but Amrit's mother immediately told me that I should not be there and sent me out on the porch to wait.

I stayed on the narrow porch—leaning against the whitewashed mud wall, watching the children play, and trying to imagine what the chanting coming from behind the door might mean—until the priest finished and called me into the house. He told me to turn on my tape recorder. He then took the recorder in his hands, pressed the microphone against his mouth, and began his chant. Amrit's mother became agitated. She said that I could ask anything I wanted, but should not try to talk to the priests and shamans. The spirits came to the healers in their dreams, Ganesh's mother, who was there as well, added. This was dangerous knowledge. I was not strong enough to receive it and might die if I did.

The following week, Ganesh's sister Padma came into my room and, with no introduction or greeting, said that Roland had died. Roland was a linguist who had spent two months in Hedangna in 1990 studying the Yamphu language. Surprised, I asked how she had heard about his death. With the same authority she had used when she once described to me the details of her wedding, she repeated what she had said and then explained that Chute Rai had told her mother that Roland had died. No one had heard from Roland in the past year. Even if something had happened to him, it made no sense that Chute would have been the villager to get the news.

Padma then asked to hear Kharka Bahadur's chant that I had recorded the previous week. I realized that her mother believed that Roland had died because he also had studied the priests' chants. (Roland later told me that he had not.) I turned on the recording. Padma clicked her tongue disapprovingly as she listened. She told me to turn it off and then said that the chants called the ancestors. I could ask the priests what they did in rituals and what the rituals were for, but I should not ask them to tell me the words they used to summon the ancestors. I explained that Kharka Bahadur had recorded the chants on his own accord. Ignoring my explanation, she said that I was a different caste. She was not sure I should have this chant or even if I should listen to it. She suggested that I tell someone older, like Chute Rai or Jadu Prasad, that I had this tape and ask him what to do.

Several days after this incident, Jaisita told me that she had met Purnamba, another priest in the village, on the trail. He told her I had visited his house that morning to ask about his work as a priest and then instructed: "Tell her not to ask about such things. If I explain them to her, I will get sick, my guru [teacher, but in this case, the spirit who teaches healers their chants and possesses them during their journeys into the realm of the ancestors] will become angry with me and will stop coming when I call."

◊

I heeded their concerns and turned my attention to questions about land disputes and local politics—issues I initially had feared would be sensitive. I started by asking older men about the history of the village and, especially, about the origins of *kipat*. What I found was that these men loved to talk about anything related to the past, especially *kipat*.

Kipat is a Nepali word that refers to individual plots of land, but it means more than that. *Kipat*, the men explained, is land cleared by their ancestors and passed along; land they do not have to buy. It means "old things" and connects them to a past that was more glorious than that of other ethnic groups in Nepal, to a time hundreds of years earlier when their Kiranti ancestors were so powerful that the king of Nepal had to strike a deal with them, swearing on a royal decree written on copper that "until cats grow horns, until the snow on the mountains melts, and the water in the rivers run dry," they would be king on this land.

After repeating the story of how Minaba and Sepa, the sons of Yamphuan, found their way to Hedangna, the men told me another. After selecting Hedangna as their home, Minaba and Sepa had to decide where each would settle. Each brother went into the forest to stake his claim. Sepa found a durable hard wood for stakes to mark the borders of his claim. The wood was strong, but looked freshly cut for some time after it had been felled. Minaba used wood that was not nearly so durable, but dried quickly and looked old even though it had just been cut. He put his stakes next to those pounded in by Sepa.

After marking their boundaries, the brothers met at a small lake on the ridge. They sat down to eat a snack. Minaba had brought pounded rice; Sepa had roasted corn flour. Sepa put a handful of dry flour into his mouth. Just then, Minaba asked, "Where is the land you have selected?" His mouth stuffed with flour and so unable to speak, Sepa waved his hand across the land and grunted. Minaba said that he could not understand his brother's gesture or his grunt. As he was eating only pounded rice, Minaba was able to speak clearly. He pointed to the gentle slope on the eastern side of the ridge, the land that is now Hedangna, and said, so his brother could understand, "This is the land I want."

Sepa protested, claiming that was the land he had chosen and that his stakes marked those borders. "So have I!" Minaba exclaimed, and the two brothers set off to examine the evidence. On arriving at the posts, placed side by side—one old and dry, the other new and fresh—Minaba said, "Ahh, you see, my posts are older; they were here before yours. Clearly, this land is mine." Although Sepa had placed his posts first, he had no proof. He was unable to dispute Minaba's claim to the land. He sighed, finished the last of his flour, lifted his small sack of belongings, and set off to settle in the west, where the land was steeper, drier, and rockier.

◊

What immediately struck me on hearing this story was that even eleven generations ago, when there were only two people in their region, the Yamphu emphasized the necessity of staking a claim to the land. And even then, when there was so much land to choose from, they disagreed. This was interesting in itself, but it was not until I conducted interviews about land disputes occurring over the past hundred years in the village that I began to see the story's relevance to *kipat*.

For the Yamphu, the king's promise that they would be king on their own lands meant that he could not "touch" land held as *kipat*. He could tax his subjects, but not their land, and he could not survey—or make any legal record of—the landholdings in the village. From the perspective of the central government, land in Hedangna was legally invisible.

In concrete terms, this meant that until the land was surveyed in 1994—a year after I lived in Hedangna—boundaries to fields were not marked, and so there was no written record of where one landowner's fields ended and another's began. Divisions between plots and decisions about who had what rights to these plots were ambiguous and negotiable. Written documents were needed only where neighbors might encroach on each other's land, either because the land was particularly valuable or because neighbors and relatives could not be trusted. Where there was trust, the demarcation could be less clear.

Disputes were the main way of making boundaries between fields visible and attaching them to particular landmarks. The Yamphu of Hedangna were renowned in the upper Arun Valley for the length and frequency of their disputes. They spanned generations; sons inherited the anger and the evidence. An older man told me that he had learned how to read during the years his father lived in the district center, pressing his land claim in court. When I asked a twenty-five-year-old man about the division of labor between men and women, he said that men were responsible for plowing fields and attending to land disputes. "Women can't do disputes," he said. "They don't know how to talk."

Scratch the surface of Hedangna's history, and it seemed as though the history of the village were a story of one long struggle among the residents to steal the land of a neighbor: a neighboring household, a neighboring relative, or a neighboring village. Villagers used the words *batho* and *sojo* to distinguish between those who started disputes and those who did not. *Sojo* is translated as "honesty," but it implies an openness, a lack of skill in asserting one's own interests, a kind of stupidity. *Sojo* people rarely initiate disputes. Those who do are considered *batho,* or clever and skillful with words. I once asked an older woman whether being *batho* was good or bad. *Batho* is good for you, she answered, and bad for everyone else.

By asking about these disputes, I found that I was able to gain a more accurate and nuanced understanding of the different ideas the villagers had about the land than if I asked about those ideas directly. In historical interviews, the details the men chose to include or exclude were dictated by my questions; their memories; and, perhaps most important, their desire for events to have happened in a particular way. Their stories emphasized the grandeur of the Kiranti, the ancestors as "kings on their own lands," and the power and independence they had enjoyed. While interesting, the stories let me see how the villagers liked to imagine their past, not how that past had actually unfolded. Talking about disputes became a way of getting beneath these stories, of seeing the range of visions and experiences encapsulated in the narrative the men told about the community of *kipatiya* (*kipat* holders).

◊

One morning, I walked along the main trail to the blacksmith's house, just north of Hedangna. It was early, the sun had just angled over the ridge that rose to the east across the Arun River, and I pulled my shawl more closely around my shoulders. I climbed through the terraced fields that would soon be planted in rice and rounded the bend past the prayer wall. Harka, a young man in the village and Raj Kumar's cousin, walked by, bent under a large load of six-foot-long posts. Harka was short and, like most men and women in Hedangna, had strong, stocky legs. He was shouting as he walked. He stopped at the house I had just passed, threw the posts on the ground, and turned immediately to yell a response at a voice coming from a house across the gully. I continued to walk and turned into the courtyard from which the other voice was coming—the home of the blacksmith.

Ganga, a thin older man who was the poorest and, people said, the laziest man in the village, stood at the edge of the mud courtyard, shaking his fist and shouting back toward Harka.

A few minutes later, Harka stomped back into the courtyard, brandishing one of the posts. Harka's father had died when he was ten. He had been forced to drop out of school to care for his family's cattle and water buffalo. Like many villagers who were unable to attend school, Harka had taught himself to read and write from schoolbooks he took with him to the pastures where he tended the livestock. Now in his early thirties, Harka was considered one of the harder workers and more innovative farmers in the village. I had never seen him so angry.

He shook the post in Ganga's face. "Is this a fodder tree?" He turned to Ganesh, in whose home I lived and who had come to the blacksmith's shop to have his sickle sharpened: "Everyone knows this type of tree is for fodder! These are my trees! I got these seedlings from the nursery. I planted them myself so I wouldn't have to climb the ridge every day to collect fodder!" He pointed toward a clump of trees in the gully beneath his house. "Go look!" he called out to me. "Walk down there and see if there aren't fodder seedlings planted between the cardamom and bamboo!" He raised the post against the deep-blue morning sky and shouted again at Ganga: "I planted this tree! I raised it from when it was small. I watched it grow year after year and I became attached to it! I felt *maya* [love, attachment] for it. And now it is cut! What can I do? In all the years it took me to raise this seedling, in one minute it is cut! Killed! What can I do? Nothing!"

Ganga, a scraggly mustache growing over his mouth, had backed up and was now squatting on the porch of the blacksmith's house. He shouted back, "Don't say those are your trees! Don't say it's your *kipat!* It is the community's forest. That wood is for everyone to use; no one has his own forest!"

Harka looked at Ganesh in despair. He dropped his fists to his sides and paced back and forth in the courtyard. He appealed to Ganesh, to the blacksmith, to me for support. "Of course people have their own forests!" he cried. "Ganesh, your forest is up the hill, right? Dilli's is over there; my forest is up the trail toward Bakle. It is part of our *kipat*, the *kipat* we inherited from our fathers. Go to that forest if you want! Cut six loads of posts from the jungle there if you want—go ahead! But don't come near my house, where I have planted fodder seedlings. This land is for planting crops! Any tree can be used for posts. Don't use my fodder trees for fence posts!" He held up the post again: "If this had been left to grow, in a few years anyone who wanted could have come to cut branches for fodder. But now the trees are finished. There isn't anything left."

Ganga shouted back a bit more, disputing Harka's claim that the trees were his, asking for his proof of ownership, but his voice lacked the conviction and passion of Harka's. They argued a bit more about what evidence was needed to claim the land, about what would happen when the government officials came to survey the land for the first time in a hundred years. Harka seemed to calm down a bit. Then he suddenly remembered what had happened. He again dropped his arms in despair and repeated that the trees had been cut and there was nothing he could do.

He began to shout again at Ganga, who was crouching under the porch roof: "I could go to Khandbari and complain. If it were anyone else, I could object. You would be fined, and at least I would be compensated. But you are my relative, and I can't do that."

He turned to me and remarked that last year, Ganga had stolen a load of fence posts as well. He also had not objected at the Forestry Office in Khandbari, but only had yelled at Ganga: "This year, when I see him cutting these trees that I planted, anger rises inside of me. But what can I do? He is my brother. We are of the same bone. I can't do anything!"

When I later spoke with Harka about the argument, he reiterated that he had not complained about Ganga because he was his relative. "Plus," he added, "Ganga's father had twenty-five *muri* [measurement of land] of rice fields, and Ganga sold all of it to pay off his debts." Ganga now lived in a ramshackle, one-room bamboo shack with his wife and daughter. They no longer had any land of their own, but got food by working as day laborers in wealthy villagers' fields.

"What can you do with a person like that?" Harka asked with exasperation.

◊

Because claims to *kipat* land were oral rather than written, the strength of one's claim depended more on the persuasiveness of the parties involved and the cleverness with which they constructed their evidence than on what actually happened or whose evidence was more legally valid. Minaba had used the quick-weathering wood to strengthen his shaky claim to land. More recently, other strategies were employed: stones and ridges were renamed, documents created, and new alliances forged by means of land transfers or marriage—all in an effort to spin the most compelling tale and to convince others that you had the power to make your story stick.

The building blocks for the disputes were old documents kept in locked, dust-covered boxes and stored in an upstairs corner of people's homes. Every household had an almost identical box. Sometimes when I sat talking with older men on their mud porches, they would bring out these old boxes and unlock them with tiny keys that hung on strings they wore around their necks. When the boxes opened, yellowed documents rolled into tight bundles and tied with string spilled onto the porch. The men searched for a particular one they wanted to show me, untying bits of string and unrolling it. Line after line of old Nepali words written in black ink stretched down these long scrolls of handmade paper. Some documents were three feet long and cracked with age; water stains had smudged whole paragraphs, and edges had been chewed by moths—words eaten and lost forever.

Although from the perspective of the government *kipat* was invisible, everyone in Hedangna knew whose land was whose. Most landholders could name the owners of their fields back four or five generations, and headmen knew which fields were part of whose *kipat*, which had been acquired in another way, and who the original owners had been. Regulations introduced in the Land Acts of 1964 required *kipatiya* to list their fields in the Land Reform Office in Khandbari, and the Yamphu, who had not paid a land tax in the past, now had to pay 5 rupees per field. Even after 1964, land that had been transferred to a person who had a different headman than the previous owner was still considered part of the original landowner's *kipat*. Taxes on these fields were paid to the headman of the former owner, not the current owner. Thus in

1992 and 1993, some individuals paid taxes to two or three headmen. Likewise, a headman collected taxes from his own subjects as well as from subjects who held land that had been included in the 1894 registration for the headman's predecessor.

Thus from the perspective of villagers, the flow of taxes reenacted the history of a plot of land, following a network of relationships that was stored in the memories of the headmen and their subjects, not in the wooden cabinets at the district tax office. Although in practice *kipat* land could be bought and sold like land held in any other system of tenure, selling it did not sever the land from its history. The sale just added another layer to that history. While this is true of any sale of titled land, because of the way land rights were staked under *kipat*, securing a claim to that land depended on remembering—or contesting—this history. The past was no more fixed than were the boundaries on the land.

Because boundaries of landholdings were not marked physically or legally, manipulating the gap between what was claimed orally and what was documented legally was one of the primary ways for villagers to demarcate the land and increase their holdings. The victor in a land dispute temporarily secured his or her (though usually his) claim to the land until the borders were redrawn as the result of another dispute. The better the land, the more likely a dispute would arise and thus the greater the need for a plausible story to support a claim. The better the story, the stronger the boundary. The importance of disputes as a way to secure land—the role of stories in making boundaries—was in large part caused by the ambiguity and complexity inherent in the *kipat* system.

Yet stories that acknowledged that ambiguity and complexity—ones that made them visible—rarely won disputes. The types of stories told to stake claims to *kipat* land depended on denying the ambiguity that the *kipat* system itself created and perpetuated. This confused me for a long time. I came to Hedangna looking for stories, believing, naively perhaps, as Barry Lopez has written, that "stories have a way of taking care of us" (1998:48), especially stories told by indigenous people. The stories I was discovering in Hedangna seemed to take care of some people at the expense of others. They were more about obtaining power, in the form of land, than about providing care and sustenance. They seemed to create more distance, not less.

Although older men described *kipat* as a coherent force in the village and emphasized the community of *kipatiya,* as I asked about their perspectives on the tradition of land disputes, I began to see that the reality of life in a community of *kipat* holders was not as I had envisioned it from afar nor, more generally, did the community in Hedangna resemble my definition of the word "community." For years, I resisted accepting what this really said about this system of tenure and about common property in general. I kept trying to understand the significance of *kipat* in terms of its content, believing that there must be something redeeming about it, something that would allow me to underscore the cultural value of a system of tenure that was more open-ended and flexible than private property.

◊

On a cold day in November, several years after I had left Hedangna, Brian and I visited the Nez Perce reservation in Idaho. Black Beaver, a Nez Perce with a long dark braid, showed us around the wolf rehabilitation center. We crouched beside the chain-link fence and looked into the eyes of the wolves, listening as Black Beaver described how two or three hundred years ago, all that had protected his people from the wolves was the thin buffalo skin of the tepee walls. Even though they had little physical protection, he explained, his ancestors could coexist with the wolves because they paid attention to the wolves' lives. They watched when the wolves bred and raised their young, watched how their behavior changed through the seasons and the cycles of their lives. This intimate understanding of the wolves helped the Indians know when the animals might be dangerous and when humans could safely enter their territory.

Black Beaver then described the fall from this state of grace, triggered, as always, by the arrival of the white settlers. The whites, Black Beaver said, built wood houses and put up barbed-wire fences to protect themselves from the wilderness they had entered. Unlike the Nez Perce, they did not pay attention to what the wolves were doing. The whites feared what they did not understand, he said, and so they needed guns, walls, and fences for protection.

I thought about what Black Beaver said in terms of *kipat*. His story was powerful to me not because it offered a vision of how the Nez Perce once lived in harmony with the wolves. Nor, I finally realized, was *kipat* important because it offered an inherently good way of regulating rights to the land. Nothing was so simple. The story about wolves—and the system of *kipat*—are important because they provide insight into the challenges of living in community with the human and natural worlds, of being in *relationship* with those worlds, not trying to control them. Black Beaver's story touched on what is needed inside to be able to step outside William Cronon's black box and be transformed by what you encounter—what it takes to live in a world that you cannot control.

In Hedangna, the objective of a dispute was to attach one's own boundaries to the land. Yet the Yamphu seemed to know that these lines were gestures at controlling what could never be controlled permanently, at fixing what could never be fixed. Neighbors would continue to encroach on each other's fields. Under *kipat,* strong and secure divisions depended on strong and secure relationships with the neighboring landowners, not on the impermeability or rigidity of the boundaries themselves. The relationship was what the villagers could count on, not the fence or the legal institution responsible for keeping that fence in place.

I finally began to understand that *kipat* was important because of the attitude—the relationship with oneself and with others—that it engendered, not because of the details of its administration. If neighbors could not be trusted to respect each other's rights to land—trusted not to cut down each other's trees in the night or plow each other's fields—they had to appeal to a third party to spell out and uphold their rights. They needed someone else to speak for them, trusting another's voice more than their own. As long as they could count on that institution or individual for protection, they were relatively secure. Yet, in exchange for that security, they lost or never acquired the knowledge needed to uphold those rights on their own.

As I became immersed in the intricacies of land tenure and the logic of the apparatus of property ownership, I began to understand how different my ideas about people and the land were from Brian's. For Brian, at the time, buying land was a way of saving it. On some level I could agree with that approach, but as I came to understand how systems of property imposed particular ways of interacting with and relating to land, I became more and more skeptical that a change in ownership was the right solution to land conservation. Ownership itself seemed to be the more fundamental problem. Like Black Beaver's story, in each of the disputes I asked about, the relationships between the landowners shaped the outcome as much as or more than the "law." The politics of *kipat* shaped the villagers' relationship with the central government. Yet something else shaped their interactions with one another and with the land. Understanding this took me away from discussions of *kipat* and into conversations and connections that I was more likely to have with women than with men.

9

GOLD EARRINGS

July 1992

A single brass kerosene lamp burned on a shelf carved out of the mud wall. The flame lit the immediate area; the rest of the room was dark. I had come to Devimaya's home for dinner and had asked Dilli Prasad, her father, some questions about *kipat* and about land disputes. I felt more comfortable with Devimaya than with anyone else in the village, so as the year went on, I began to spend more and more time with her family. Dhanmaya, Devimaya's mother, sat in the shadows by the dying coals. The dishes were cleaned. The three youngest children lay on straw mats by the fire where they had fallen asleep after eating. Dhanmaya pulled blankets over them and turned back to stare into the coals.

Dilli sat cross-legged on the far side of the fire. He was telling me about the Golechaur land dispute, in which he had been involved for years. I had been hearing about land disputes all week and was more curious about Dhanmaya, wondering why she was so quiet. I wondered if she bothered listening to these conversations about disputes, if she even was paying attention. She occasionally looked up as though following what Dilli was saying, but she rarely said anything. Then she inhaled the last bit of smoke from her home-made cigarette, stuck the lit end of the leaf onto the tip of her tongue to put it out, and stood up. Stooping under the low ceiling, she walked to a hand-carved wooden container hanging from the center post of the room, took off the lid, and scooped out a clump of *ghue* (clarified butter) with her fingers. She returned to her seat by the fire, stuck the butter onto the edge of a bronze bowl, and lifted her left foot.

The sole of her foot was the color of dirt; it was thick and calloused from a lifetime of walking barefoot across the rough lands of the upper Arun Valley. Deep cracks from the dryness and the cold cut through the calluses. I had similar cracks in my own feet and knew how painful they were, especially when walking through water.

Dhanmaya owned a pair of green tennis shoes made in China and bought by Dilli in the Khandbari bazaar, but she rarely wore them. She did not trust the rubber soles not to slip on rocks and roots, especially when she was carrying heavy loads. Plus, people would talk, she said. They would accuse her of thinking that she was superior. So she preferred to go barefoot, like everyone else.

Now she brushed the mud and dirt off her left foot, put a bit of butter onto her right hand, and pressed it into the split skin on the foot. She did this slowly, making sure that the oils penetrated into each of the cracks. After she rubbed in the grease, she raised her foot over the coals. Once the *ghue* had melted, she brought her foot back to her lap and squeezed the split skin together, holding it closed for a moment so the oils would soften and moisten the skin. She looked closely to make sure the butter was rubbed in and then repeated the entire process on the other foot.

Dhanmaya had grown up in a community across the Arun River, on the ridge to the east, the eldest of six children. The land was rougher there, but the village was smaller, and water, fodder, and firewood were closer than in Hedangna. Whenever Dhanmaya talked about the differences between the village where she had grown up and Hedangna, she would pull back her hair to reveal how her hairline had receded from the tumpline of the baskets of firewood and water she had carried since moving to Hedangna.

As I watched Dhanmaya, I tried to imagine what she had been like when she met Dilli, what life had looked like stretching out before her, young and proud and admired by one of the wealthiest men in the region. In trying to envision

Dhanmaya as a young woman, I looked first at Kumari, whose life would most closely follow her mother's.

When I met her, Kumari was fifteen. She had long arms and legs and feet that embarrassed her because they were so much longer than everyone else's. She passed her time tracking down water buffalo and goats she had been told to watch and tromping through the forest with friends to gather firewood, stinging nettle to feed to the pigs, or a particular flower to make chutney. Kumari would be sent to fetch water, a twenty-minute round trip, and would disappear for an hour or two, returning only after her mother sent one of the younger siblings to find her. Kumari did what she was told, but did it distractedly, always ready to pause for some gossip or the chance to scurry up a trail in search of *lapsi* seeds or a cricket to fry with dinner, taking twice as long as Devimaya or Dhanmaya would have.

I often beat rice with Kumari, stepping up and down for hours on the *dikki* (heavy wooden beam that dehusked the rice). Dehusking rice is tedious, tiring work. It also requires no skill. Once I figured out how long it takes to do alone and how physically demanding it is—like going for a hard two-hour run— I helped whenever I could.

Kumari and I talked to pass the time. Mid-sentence, she would duck out from the shed, squat in the dirt to pee, and then leap back in step on the wooden beam to complete the sentence where she had left off. Devimaya was more inquisitive and thoughtful than Kumari; she had a stronger sense of duty and responsibility. But Devimaya had been sent to school, given a chance at something different, whereas Kumari was needed at home. Devimaya got an education, and Kumari got heavy gold earrings and a longer gold nose ring— her inheritance, Dhanmaya explained, to make up for her inability to read.

After I moved to a small building next to Dhanmaya and Dilli's house and began to eat my meals with them, Kumari would escape to my room when her father arrived home drunk or angry. She told me that he scolded her mother for not sending Rendha, their only son, to school and for letting the children run around dirty. He berated Kumari for letting the water buffalo get into the neighbor's cardamom and yelled at them both for doing nothing all day but eating and shitting, eating and shitting. Later, over meals, Kumari made faces behind her father's back and refused to eat, rebelling in the only way she dared.

I saw Kumari when I looked at Dhanmaya, saw Kumari's spirit and rebelliousness in the way Dhanmaya held up her chin and in her refusal to complain. I also saw sorrow and exhaustion and a quality of endurance that I had not often encountered. Ten or fifteen years earlier, I wondered, would I also have seen the pain of missed opportunity and of disappointment with a husband who, like so many of the men in the upper village, drank, played cards, and had a bad temper? I did not see much of that now, nor did I hear it in her words. Mostly I saw and heard the incredible love she had for her children. I saw what it took to bear that love.

Dhanmaya had borne fourteen children, nine of whom had died, three in a measles epidemic that swept through the village in the late 1970s, wiping out too many people to count. Dilli and Dhanmaya had gone to the forest to bury one of their daughters, only to return home to discover that a son had died. Fifteen days later, another daughter died. Several years after that, their only other son died. Any disappointment with her own life was now outweighed by an almost desperate love for the five children who had survived and by the knowledge that it was only her diligent, constant watchfulness and her back-breaking work that kept these children alive and the awareness that even this work might not be enough.

Dilli once went north to the Bhotiya village of Chepuwa to collect money for an ox he had sold. He said that he would return after three days. A week went by, and then ten days, and still he had not returned or sent any word. Maybe he had taken a Bhotiya as a second wife, Dhanmaya said.

"Let him," I said, "wouldn't that be easier?"

She shook her head. "Look at me," she replied, pulling at her hair. "I have no teeth left; my hair is falling out. Where would I go if he took someone else? Who would take me?" Her husband's property fed her children, the roof of his house covered their heads, and the children belonged to him: these were the rules and customs of the village. If she left, she had nothing. And so she stayed.

◊

Another night, we were again gathered around the fire quietly eating rice. Rendha started to eat his food and then pushed away the tin plate with his hand. He wanted his food on a bronze plate, he announced, not a tin one. Dilli told Dhanmaya to get him a different plate, which she did. Rendha shoved away the bronze plate with his foot and began to cry, repeating that he wanted a bronze plate, even though he now had one. The daughters silently continued to eat what was on their plates, pretending not to notice. I did as they did.

His cries turned to shrieks, and he began to demand meat. The shrieks became louder. He threw globs of rice into the flames and again kicked the plate with his foot. Dilli told Dhanmaya to find him some meat. Seemingly at a loss for what to do to stop his crying, Dhanmaya stood up and dug through the basket hanging over the fire, frantically searching for meat. Dilli climbed the ladder to look upstairs. Rendha kept shrieking. The girls and I continued to eat. Dilli eventually returned with a string of dried yak meat. He put several pieces on Rendha's bronze plate, several on his own, and placed the rest in the basket above the fire. Rendha immediately stopped crying and gnawed the stringy meat, eyeing his sisters with glee.

The daughters did not look up. They quietly ate the rice mixed with bits of watery spinach piled on their tin plates. No one said anything. Everyone finished their food and went out to wash their hands. It was not until much later, after the food had been put away and the plates put outside to be washed, that Devimaya broke the silence. She asked, so quietly and with such hesitation that no one even acknowledged her words, "What about meat for the daughters?"

◊

Dhanmaya loved all her children. She spoiled them all. How could she not, she asked, when she had lost so many. Yet, because of the community they were born into, she—a mother who was also a daughter—spoiled the son more than the daughters. Her daughters were her closest companions and her most dependable workers. But, like all daughters in the village, they eventually would leave. Even though recent legislation gave daughters the right to inherit land, few families had enough land to ensure that their sons could support a family, let alone leave any to daughters. Women depended on their husbands' property to care for them and their children. Once married, they were welcome to return to their *maiti* (parents' home) to visit; they were rarely welcome to stay. Attention, if not affection, was inevitably attached to the future. In a community where survival was so precarious and land so limited, was there any choice?

Once when Devimaya and I were beating rice on the *dikki*, we talked about what we would most like to have or who we would want to be. With no hesitation, she said that she wished she were a boy.

◊

Another evening, Baiseti Thuma (*thuma* means "grandmother" in Yamphu) joined Dhanmaya, four of her children, and me around the fire after we had finished our meal. Devimaya was in a village near Khandbari studying for her high-school certificate, and Dilli was wandering around the village.

Baiseti Thuma lived in a one-room house next to Dilli and Dhanmaya. She was the widow of Dilli's father's brother. She had no sons of her own, so she had moved closer to Dilli, her only surviving male relative. Eleven of her children had died at birth; of the two daughters who had survived, one had died in childbirth. Her only surviving daughter lived up the hill with her husband. Baiseti Thuma was poor, having lost most of her money and land in a viscous dispute with her nephew, Dilli's cousin, after her husband died.

Baiseti's head barely came to my shoulder; her wrists were as tiny as those of a small child; and her skin was deeply wrinkled and dark from years of living and working in the sun, the cold, and the heat. She had a chronic cough from smoking too much, and she laughed more than anyone else I knew in the village. She could not afford kerosene, so I gave her candles and she told me stories about the ancestors who had roamed the fields and forests, stories told in a tired, raspy voice that I could barely understand. Baiseti Thuma was a shamani, Devimaya had told me; she had visions and dreams like those of the shamans and priests in the village. When I asked Baiseti, she said that it was true, but she was too shy to perform for others; plus, she added, she was a woman, and women did not have time for that sort of thing.

Shadows cast by the fire flickered across the soot-covered ceiling and danced across the grandmother's smiling face, one or two stained teeth still remaining in her mouth. Her gnarled dark hand reaching up to position the pale-blue, threadbare shawl on her head, she began to describe to Kumari how she must behave in her future husband's family. Kumari protested that she was only fifteen, too young to think about getting married. Whether or not that was true (Kumari ran off with a boy to India two years later), it did not stop Baiseti.

Baiseti Thuma first demonstrated the proper way to sit in one's in-laws' house: knees together and bent under her, bottom resting on her feet. She then imitated how Kumari sat: feet together, knees splayed to either side, bottom in the air. Baiseti stared into space, mouth open, and swept her hand across the floor; without looking, she picked up whatever she touched and put it into her mouth. As she did this, she gave a running explanation of her actions, speaking in a voice so raspy that it was difficult to hear. Then she sat back down, repositioned the threadbare shawl on her head, and in a much more efficient tone said, "Now, Kumari, this is how you must sit when you go to your husband's house." She again demonstrated the correct way to sit: knees together and bent under her, bottom resting on her feet. "If you squat, with your bottom in the air," she said, "gas will slip out." She imitated the sound that gas would make. "If you let that happen," she concluded, "you will make your parents feel shame."

Kumari, her hair sticking out from when she had been feeling around for lice, was sitting with her feet together and her knees dropping down on either side, just as the grandmother had been telling her not to sit. "Keep your knees together," Baiseti told her again. Kumari tried it, said that it hurt and she could not sit like that, and went back to sitting as she had been.

I loved Baiseti, loved her more than I ever loved my real grandmothers. I went to her house often and always felt more peaceful and centered when I left, even though I was able to understand only a portion of the stories she told me, her voice was so raspy.

Baiseti Thuma had been married off when she was so young that she had to be carried to her new home in a bamboo basket. She then lived for two years with her in-laws without saying a word. During one of my visits to her house, she made me some *jad,* which she rarely did because she did not have enough for herself. While I drank, she described having gone to a church when she was visiting her sister in Kalimpong, in northern India. Baiseti crouched low, hands pressed together before her heart and eyes squeezed shut, to show how the congregants had prayed. Then abruptly she rose, head and arms thrown back, to demonstrate how they began to sing. Just as abruptly, she returned to the crouched position for the prayers. And up again to sing. When she finally sat down and we stopped laughing, I asked why she had not stayed with her sister in Kalimpong, where life must have been so much easier. Because she had two daughters, she told me, and she missed them all the time. And her husband had died in Hedangna, so she wanted to die here as well.

Whenever I returned to the village, Baiseti came to Dilli and Dhanmaya's house, clutched my hands, and began to chant a rhythmic beat, dancing with me in a circle around the mud courtyard. Other times, especially on the gray rainy days of winter or the hard monsoon days of summer rice planting, I would pass her as she squatted on her porch. While she rolled some tobacco in a dried leaf to smoke, she would tell me that she did not want anything anymore, did not want clothing or money; that life was too hard. She had lived too long, she would say, and just wanted to die.

Finished with the topic of Kumari's sitting habits, the grandmother moved on to a description of a recent dream. She told us that she had awoken in the middle of the night, lips smacking, looking for a cigarette. She showed us how her lips had been smacking, her eyes squeezed shut to remind us that she was asleep. "Dhanmaya tells me that I must stop smoking," the *thuma* said, turning to me, "that that is why my voice is so rough. How can someone who dreams like this," she continued, eyes shut and lips smacking, "think about giving up smoking?"

◊

March 1997

On my last trip to Hedangna, in 1997, Devimaya and I went to Uwa, a village just north of Hedangna, to visit her aunt. The morning before we returned to Hedangna, I went next door to express my condolences to Chakra Bahadur, an older man who had helped me a tremendous amount during my previous stay in the village and whose youngest son, Dev Kumar, had recently fallen from a cliff and died. Dev Kumar had been Devimaya's fiancé. I told Chakra how sorry I was, how terribly, terribly sorry. He nodded, went inside his house, and came out with a photograph I had sent of Chakra and Dev Kumar standing stiffly in front of their house. He told me how much this photograph meant to him now that his son was gone. He was broken now, he said, shaking his head. He turned and went back inside.

As I turned to leave, I saw Aulimaya, a young woman I had known from Hedangna, sweeping hay and clumps of dirt from the mud courtyard just below Chakra's house. She waved and gestured for me to follow her into the small hut below Chakra's house where she lived with her husband, Chakra's eldest son. Yamphu women tend to be bold and loud, with a colorful sense of humor, but Aulimaya was quieter than most women in the village. She was beautiful and thin and shy, and I had liked her from the first time we met, a day or two after I had moved into the room on the porch of Ganesh and Jaisita's house. I had been struck by a sad, almost haunted look in her eyes, as though she longed for something she could not express. Our paths did not often cross, and I was grateful when they did.

Aulimaya pulled out two small mats and offered me some *jad*. The coals were out, and the house looked bare. I could tell that she was being polite and said that I had just drunk some and my stomach was full. She nodded and pulled her mat close to mine. We sat cross-legged in the cool mud-floored room, listening to the high-pitched peeping of the chicks. She shooed away the children who had come to listen to our conversation and talked about Dev Kumar.

Although close in age, Devimaya and Aulimaya had never been good friends, because Devimaya had attended school and Aulimaya had not. After Aulimaya married Chakra's eldest son, though, Dev Kumar would secretly ask her to carry letters to Devimaya when Aulimaya went to her *maiti* in Hedangna, and Devimaya would ask her to take letters to Dev Kumar on her return to Uwa.

For years, Aulimaya counted on having Devimaya as a sister-in-law sometime in the future. Although Devimaya and Dev Kumar hoped to get jobs as teachers in the Hedangna school and would likely have rented a room in Hedangna, whenever they came to Dev Kumar's house, Aulimaya and Devimaya would have been neighbors; they would have shared the work that Aulimaya now did on her own or with her mother-in-law. She would have had a friend who was also a sister-in-law. Aulimaya had lost not only Dev Kumar, who had become her friend, but also Devimaya.

Aulimaya still dreamed of him, she said, still imagined that one day he would walk through the door, pull up a mat, and sit down to talk. He always made her laugh, and she missed him very much. His mother did nothing but cry all day; that was her work now: crying. Aulimaya quickly looked down at her lap, where her fingers pulled at the threads in her shawl. Abruptly, she again shooed away the children.

For the first month, all she could think about was Dev Kumar's death. And if she felt so sad, as though she wanted to cry all the time, how must it have been, still be, for his parents. She had not had a child; her parents were still alive; she could not understand how Dev Kumar's parents must feel. Nor did she know what it was like to love someone and lose him, as had Devimaya.

"I'm not in love," Aulimaya said, abruptly changing the subject. "I've never been in love." Aulimaya had not even seen her husband before she was married to him. Chakra's oldest son, now her husband, had needed a wife. Some people had asked around for recommendations of marriageable women who had good habits and worked hard, and then Chakra's relatives had brought *raksi* to her father's house. Her parents could accept the *raksi* and drink it, which signaled that they had agreed to the marriage; return it and wait for someone else; or let their daughter choose on her own. She had watched from inside the house as her parents and Chakra's relatives talked, and then she saw her parents accept and drink the *raksi*. That was that. She had been married to a man she had never seen. And she had not even been asked.

She moved into Chakra's house and did not speak to her husband for a year and a half. He would call her, but she never answered. Aulimaya shook her head again and repeated that Chakra's mother did nothing but cry. Again she looked down at her hands folded in her lap.

When she looked up, she asked, "When do you have to fuck your husband to have a child?" I use that word because that is the Nepali word, *cicnu*. Women never talked about "making love," which in Nepali is *maya garnu*. Older women who had been married for a while dismissed the idea that marriage had anything to do with love. This was not always true, especially with the "love" (as opposed to "arranged") marriages that were becoming more common.

Aulimaya's family was not wealthy; she had several brothers who would divide the land they inherited; and she had not had a choice once her parents decided that she was to be married. The word she meant now, it seemed clear, was "fuck," specifically with the goal of producing offspring.

She told me that at first she had not worried about not having a child. Her husband did not seem worried either; he had not taken another wife, which men often did when their wives did not bear children or produced daughters but no sons. She saw how much trouble it was for other people to have children, and she thought it was just fine without them. "I'm still not worried," she insisted. "I'll worry only in a few years when I get too old. Now I'm only twenty-six or so, and it is still okay."

But then she kept talking and asking questions with an earnestness that made it clear how worried she was, and she spoke softly, with an aura of secrecy, which made me realize how alone she was with her fears. She told me that when she had begun to menstruate, it would last from Friday to Friday, but now it lasted for only one day. Her back and legs hurt, her womb hurt, and she wondered if something inside was broken. "Do you know?" she asked. "Do you think something is broken?"

I asked whom she had talked to about it and whether she had been to a health clinic. She had not spoken with anyone other than me, she answered, especially not to her husband. She had not been to Khandbari or even to the health post four hours away in Seduwa.

She kept talking. She told me that several years earlier, her mother-in-law had ordered her to leave, to return to her parents' home because she had no children. At the same time, she now confessed, her husband had been living with a Gurung woman in a village across the river. She had not talked to him for two months. Then he returned. From what I could tell, they never discussed why he had left or why he had returned. He just came back. And they again began trying to have a child.

I tried to explain how to count from the first day of her period, so she would know when to have sex. I could see from her expression that what I said made no sense to her. "Go to the health clinic," I said. "They can help. Maybe it is your husband's fault. Maybe he can't have children. He has to go to the clinic as well." She nodded, but I knew that she would never go.

Aulimaya then said that I was like her, that I also did not have children. "We're both *sukeko* [dried up]." I then realized that perhaps she had decided to talk with me not because I might be able to help, not even because I was an outsider and would not tell anyone else what she had said. She spoke with me because, like her, I had no children, even though I had been married for at least four years. I was safe in a way that others were not.

While in Hedangna, I often wondered what loneliness meant for those who grew up and grew old among neighbors who were cousins and uncles and grandmothers. While talking with Aulimaya, I encountered a different sort of loneliness, not just the loneliness of Dhanmaya when she had to do the work her sons would have done, but the loneliness of Aulimaya and of Devimaya after Dev Kumar's death—the loneliness of not being able to share openly what is inside; the inability to be different; and the need to pretend, if you are, that you are not.

◊

Once I was heading through the village to speak with an older man about some disputes that had occurred on his land. A woman I knew was planting millet in her fields, so I stopped to help. We planted for a while, and then took a break to sit on a boulder nearby. She reached into the neck of her faded, dirty pink shirt and pulled out a small, worn bag tied to a sash inside the shirt. All women had such a bag for carrying money, tobacco, and keys. She took out a bit of cotton, a piece of flint, a small rock, and some tobacco; placed them in a pile on her lap; and reached back into the top of her shirt for a dried leaf.

While she began working to get a spark from the flint to light the cotton and, from that, the leaf and tobacco, I reached into my woven shoulder bag crammed with notebooks, pens, and a water bottle and pulled out a thin box of matches I had picked up at a restaurant in New York before leaving the United States. I opened the box, pulled out a match, and struck it as she again tried to get the bit of cotton to light from the spark. We burst out laughing. She held up her grimy little bag and said, "My country"; then she held up my box of matches and said, "Yours."

◊

A professor of English at Dartmouth College once said that it is not differences that divide people, but silence. I think about the different kinds of and different reasons for silence. There is the silence of what is forbidden to say: voices that are silenced for political, economic, social, and cultural reasons and that disciplines such as anthropology try to help be heard. And there is the silence of indifference, the inability or the refusal to tell the stories that connect across our differences. This kind of complacency often silences me, believing that there is nothing I can do or say that will make a difference, so why bother trying. These are the silences that divide. These are the silences that knowledge and a powerful vision can transform.

What about the silence of what cannot be said? What does it take to keep writing, keep talking, keep reaching our hands across the borders that divide, not to erase or even to transcend the differences, but to find ways to bring the silence that is bigger than words into stories created with words?

I thought back to William Cronon's black box. I assumed that the friendships I was forming with Devimaya, Dhanmaya, and Baiseti Thuma—like the work I did in the fields—were steps on the path of the research I had come to Nepal to conduct. It was not until later that I realized that, in fact, these connections and experiences were the point, that they were awakening in me the transformation needed to begin to understand the lives and culture of the residents of Hedangna and to grasp how they were different from my own. I was so focused on understanding *kipat* and land tenure, on discovering the right vision for protecting the villagers' land and culture, that I did not see that my relationships with them, the way I began to care for them and they for me, were the sparks that connected across the boxes that divided. Without that connection—without that love—knowledge was simply power and a vision was nothing but arrogance.

10

THIN PLACES

August 1992

My skin was like wet tissue paper. It peeled off with my socks, pulled off under the damp bandage. It came off from between my toes, from the soles of my feet, and from the edges of my heels. The exposed new skin was raw and tender. There was too much of it to cover and nothing solid or dry to hold down a new bandage. I had never seen anything like it and had no idea what to do. I glanced up with despair and saw the women already lifting their bamboo baskets and filing barefoot into the early-morning mist. Pain was preferable to abandonment. Wincing, I pulled on my last pair of dry socks and laced up my soggy boots. I stuffed my jacket into the top of my pack and followed in the direction of the women—day number two on our pilgrimage to Khembalung.

Khembalung refers to several places. It is Makalu, the fifth highest mountain in the world and the home, so the villagers in Hedangna say, of Lord Shiva.

At night in their dreams, Yamphu shamans and priests say, they travel to a cold, clear lake on the right shoulder of Khembalung. Witches travel to the ridge as well, but they bathe instead in a lake of blood. When they are done washing, the shamans and witches and priests stretch out on the rocks, drying themselves in the moonlight and arguing about who is the most powerful.

Khembalung is also a *bhayul* (hidden valley) of Tibetan cosmology, a pure enchanted land set outside the destruction and corruption of time. Here, so the legends say, one will find refuge from the enemies of religion and attain eternal youth, beauty, strength, and fertility.

The *bhayuls* are physical places hidden deep in the Himalayas and rendered inaccessible by the magic of the Tibetan yogi Devimayasambhava. The prophesies say that when all temples are destroyed, when servants become masters, and when "people sacrifice their own animals, drink blood and eat flesh of their own fathers," those disciples of Devimayasambhava who "display greatness of heart" will retrieve the guidebooks hidden thousands of years earlier and set out on journeys to "open" these hidden lands (trans. Orofino 1991:257).

What pilgrims see on their journey to these sacred places depends on what they are capable of seeing. Some encounter rocks, trees, and mountains covered with snow. Others who cross the same terrain see mysterious landscapes shimmering with jewels and awesome mountains floating above clouds of light.

What they experience when they arrive in the valley of Khembalung also depends on what they are ready to experience. Most find a peaceful and fertile land with room for five hundred people to settle. They receive a blessing of good health, long life, fertility, and strength; their desires will be fulfilled. Yogis who enter the same lush valley have sudden visions of the nature of reality, flashes of insight that are fleeting, but that strengthen and deepen their spiritual journeys.

Entrance to the innermost realm of Khembalung is reserved for those who have reached the highest level of spiritual fulfillment. The physical landscape corresponds to the body and mind of the pilgrim; there are no boundaries between the self and the world. When travelers enter this place—a place, so the guidebooks say, where there is no distinction between life and death—they acquire the clarity of mind and openness of heart needed to attain enlightenment.

Few attempt to undertake journeys to these hidden lands. It is too dangerous, and they fear that they will never return. But many make pilgrimages to the places believed to be "gateways" to these valleys. Two caves carved out of a granite cliff 1,000 feet above the high-altitude summer pastures of Yangle Meadow, a day's walk south of the base of Makalu, are said to be entrances into the hidden valley of Khembalung. Whether they are or not, these caves are believed to be sacred places where gods have stayed and are among the most important pilgrimage sites for Hindus and Buddhists in the upper Arun Valley.

Priests and shamans, lamas and yogis are able to journey to Khembalung in their dreams or through intense spiritual practice, but everyone else must get there on foot. And so at the height of the monsoon, I set out at dawn with twenty-five villagers from Hedangna on the annual pilgrimage to the Khembalung caves at the time of the full moon.

The Yamphu with whom I traveled knew that the caves are connected to the hidden valleys of Tibetan cosmology, but they described themselves as Hindu and referred to the site as Shiva's cave. Two Brahmans from a less remote village to the south joined us. These men were tall and thin and not at all suited to the long, hard days of walking. One of them complained incessantly that the trail was too hard and the trip too difficult. Each time the Brahman complained, Jadu Prasad, one of the oldest Yamphu men in the group, who was on his sixth pilgrimage to Khembalung, replied matter-of-factly, "It wouldn't be a pilgrimage if it wasn't difficult." By the end of the trip, we were all repeating, sometimes with laughter, sometimes without, "It wouldn't be a pilgrimage if it wasn't difficult."

While yogis and lamas journey to hidden valleys to escape samsara (cycle of death and rebirth) and attain eternal bliss, the pilgrims with whom I traveled had more modest goals. The men and women were going to Khembalung to ask for a son, a daughter, a job, a good harvest. I hoped to learn more about their pilgrimages, about what they did and why. But, like the others, I soon began to talk less as we walked. Like them, I found myself mumbling repetitive chants, over and over, to keep myself moving across the rocky terrain.

I carried a down sleeping bag, a Thermarest, a toothbrush, a pack cover, a flashlight, a notebook, iodine, four pairs of socks, long underwear, a Synchilla jacket, a camera, and rice. My boots were soggy, my socks were soggy, and the wet, soft skin on my feet continued to peel. There was nothing I could do, so I simply ignored the pain as best I could. Every time I unpacked and repacked, the women gathered around to comment on each item I had brought. Each of them carried a hand-woven woolen blanket, a bamboo mat to keep out the rain, rice, some spices, and a pot. All were barefoot. They had bundles of string and bits of cloth, shawls, their finest clothing to wear on the day we climbed to the caves, and *raksi*. That was all.

Each morning, we awoke in the dark. We walked all day along steep narrow trails, fording icy streams that overflowed from the monsoon rains and climbing from 5,000 feet in Hedangna over two 16,000-foot passes and up the Barun Valley. In five days, we covered the same distance that I had traveled in two weeks the previous spring while trekking with members of my family.

We stopped only at dusk, when we had reached a cave large enough to hold all twenty-five of us. We ate one meal a day, of rice mixed with wild plants gathered along the trail. While hiking, we snacked on roasted corn flour. Occasionally, we drank black tea.

On the third afternoon, we arrived at Yangle Meadow, the grazing pastures below the Khembalung caves. We sat on the grassy floor of the narrow valley, at 13,000 feet, flanked by towering granite cliffs. Our words were swallowed by the roar of the Barun River, which carves through the center of the valley. Jadu Prasad pointed out an invisible trail going straight up the vertical rock face: the path to the caves. I sat silently. A chill that had been with me for the entire trip slowly crept up from my stomach. The two oldest women in the group, both in their seventies, looked at the cliff and then looked at me. "Don't go," they said. "Don't do it; the trail is too hard. Stay below and wait."

I know how to rock climb, I know what to be afraid of, and I shared their concern. "If these grandmothers can do it, of course you can," Jadu Prasad said. Having spent much of the past year trying to keep up with these grandmothers while collecting firewood and stinging nettles in the woods around Hedangna, I was not so sure. But the men promised that we would all go together and that they would look out for me. If I could go with them, I agreed, I would give it a try. We lifted our loads and went in search of a dry cave in which to spend the night.

The next morning, we awoke in the dark. It was drizzling. It had rained all night, and I had slept fitfully, dreaming of slippery mud and slippery rocks. I found Jadu Prasad and asked again if he thought I could make it. Again he reassured me, so I joined the women to bathe. The women were used to doing things on their own; they were strong, and they assumed that I was equally strong. I could not count on them for help on the trail. They seemed to be taking their time bathing, and so, after a perfunctory dip in the icy water, I returned to the cave, which was empty. I waited, thinking that the men must have gone to bathe as well.

Finally, one man returned. He was surprised to see me and said that the men had left and that he had come back to retrieve something he had forgotten. I grabbed my bag and scrambled after him. We walked silently and rapidly

through the drizzle, turning off the main trail onto a narrow overgrown path
that climbed toward the cliff. We caught up with Jadu Prasad and the two
Brahmans. They greeted us as we approached and told me that the trail was
too slippery for my boots and that I should walk barefoot; they then returned
to their discussion of whether the two menstruating women in the group
should climb to the sacred caves. I was curious to hear what they had to say,
but was distracted by the trail and, now, by my bare feet. I had always imagined
the pain that comes with walking barefoot over rocks and roots. Until now, I
had never thought of the cold. The soles of my feet were numb, so numb that
I did not feel the stones underneath.

Soon the trail disappeared into the base of the rock. Those ahead had been
slowed by the climb, and the women coming from behind caught up with us.
Hands gripping the rock, we slowly followed the others up the cliff. Along
with our group of twenty-five from Hedangna, there were Bhotiyas (Buddhists) from the northern Arun Valley and Chetris (Hindus) from the south.
Together, sixty or more people were making their way up the rock face.

In the West, we climb rocks with rope and protection. We wear soft rubber
under our feet. We are on the rock, but not on the rock. With these pilgrims,
I climbed to the Khembalung caves barefoot, with no rope. Perched on a tiny
ledge, Jadu Prasad reached down to pull me over difficult sections. I clutched
his hand as he hauled me up the cliff, not letting myself think about what he, in
turn, was holding on to. At a particularly rough patch, one of the grandmothers looked at me with concern and suggested that I go down. But then a man
appeared with a twelve-foot piece of rope. He knelt above the difficult section
and held the rope as I used it to climb up the crack.

Once, at a Quaker wedding I attended, the father of the groom talked about
thin places, places where one's nerve endings are bare. People make pilgrimages to thin places, places where gods have made their mark on the land. As the
legends of the hidden valleys make clear, these journeys are internal as much
as external. What the pilgrims encounter—the blessings they perceive—
depends as much on their receptivity as on the sanctity of the land they pass
through.

Two hours after leaving the valley floor, the trail leveled and we began to
climb the final section through thick clumps of juniper. Spiky roots and sharp
stones under the juniper bushes reminded me of my bare feet, by now accus-
tomed to the cold. While climbing, we had been able to see only the rock rising immediately ahead and the valley receding below. As we came over the
last incline, the most sacred site in the upper Arun Valley—the Khembalung

caves—suddenly loomed before us: an immense amphitheater carved out of the cliff by a torrent of water pouring from an opening at the top of an enormous cave. Buddhists say that Padmasambhava meditated here on his way to Tibet. Hindus say that Shiva bathed here the evening before his wedding to Parvati. Now, snatches of the high-pitched chants of the Chetri pilgrims drifted down from the base of the amphitheater.

We approached the cave from below, first stopping at a cairn where we hung narrow pieces of colored cotton cloth. Then, in single file, we walked through the waterfall. Those before me stood directly under the cascade and drenched themselves in the freezing water. It was still drizzling and cold. Already chilled, I skirted the edges of the waterfall, hoping that no one would notice, and followed the others up the last rocky stretch and into the cave.

The air inside was cold and dry and laced with the sweet smell of burning juniper. Red and green, blue and yellow prayer flags brought by the Bhotiyas and attached to long sticks rose out of a pile of stones in the center of the cave. Smaller bits of cloth had been tied to sticks or rocks. Candles, clumps of wildflowers, red *tika* powder, coins, even a watch were placed haphazardly on the pile of stones beneath the prayer flags.

Until now, we had been quiet, focused on the trail and the destination. Once we were in the amphitheater, the atmosphere changed. There was work to be done. Two women pulled out clumps of string that they coiled into bundles, dipped in *ghue,* and lit as candles. One couple carefully placed a small tin trident below the prayer flags. A young man, who had come on the pilgrimage to assist his mother, sat off to the side, staring at the opening in the ceiling of the cave through which water flowed. There was no way that people could have made that hole, he told me. Only a god could have done so, which is why we had to give offerings. A middle-aged man who had moved to Hedangna from southern Nepal paused in his preparations to scan the amphitheater. He had heard about this place since he was young, he told me. "Now that we are finally here," he said, "we have to take our time and make sure we do things right."

The time spent in the cave was not what I think of as spiritual. There were too many people, too much commotion, too much concern about this piece of string and that piece of cloth. I was too preoccupied with how we were going to climb down the cliff. Still, the cave was awesome. Now, the voices and din echoed off the high ceiling of the cave, but I imagined being here alone, with only the sounds of the wind and of the torrent of water spraying against the rock. The ground dropped steeply, and all I could see was the Barun River, silver and silent, winding its way through the green meadows far below.

◊

We left Shiva's cave, walked down a narrow path through the juniper, and continued around the ridge to a smaller cave set in the rock face. Both caves are considered part of the sacred site known as the Khembalung caves by Hindus and Buddhists alike. For Hindus, this is the cave in which Parvati lived and is said to have bathed in the water falling from the rock above. For Buddhists, this is the cave in which Yeshe Tsogyel, Padmasambhava's consort, is said to have stayed on their way from India to Tibet. We took turns crawling into a space that could hold only three or four at a time. Light from string candles dipped in *ghue* and set on the floor illuminated exposed chunks of crystal along the inside of the cave, and the air was pungent from the burning butter. The rest of the cave was in shadow. Several red plastic bangles and a white cotton shirt sewn by a tailor in Hedangna had been placed amid the usual bits of cloth and coins: offerings brought by a couple hoping for a child.

Outside, more juniper was burned. One of the Brahmans chanted prayers for the well-being of our group, we tossed bits of uncooked rice into the smoke, and the Brahman wiped ashes on our foreheads as a *tika* (blessing). We then began the descent, taking a path that was not nearly as steep as that we had ascended. I paused to pull on my boots and followed the others to the cave in which we had slept the previous night. The menstruating women who had been prohibited, by tradition, from visiting the sacred caves sat by a smoldering, smoky fire. They added some wood to the coals to heat water for tea, and we snacked on roasted corn flour mixed with sugar. The two oldest women said that they were too tired to continue up the valley with the rest of us and that they would wait for us here. We packed our loads and set out once again.

The floor of the Barun Valley was brilliant green from the summer rains, and there was finally a bit of blue sky. The air on my bare feet that morning had dried the skin, and the raw patches felt less painful. With the climb to the caves over, I felt carefree for the first time in days. As we walked north through the valley, one man speculated that the weather had turned because of the particularly strong dharma of someone in our group. The idea that sun and rain responded to our thoughts and actions reassured me and helped me feel less alone in the vast landscape. We walked until early evening, stopping at a huge open cave at 15,000 feet to spend the night. The next morning, we hiked for a few hours to bathe and make offerings in the headwaters of the Barun River on the morning of the full moon.

◊

Another long day walking in misty rain, another night sleeping in yet another cave. There had been confusion over a bag I had left with the grandmothers, whom, we now discovered, had decided to head home before us. One of the women reprimanded me for not taking responsibility for my belongings. A man who had told me the previous day to leave the extra weight looked at me with disdain and said that he had told me he would carry the bag. I turned away and walked to the river's edge to fill my water bottle. It was dusk, and the sky still was overcast. I stood on the banks of the Barun River, alone. I thought about how hard I was trying to get it right—to walk fast enough, to say the right thing, to understand the right way. In Hedangna, I had novels to read and a tiny room with a door that I could shut, a door that, oddly enough, protected me from this stark realization of my solitude. For the past few days, these barriers had been stripped away, and this sudden and complete exposure made me acutely aware of the gap between my world and that of my companions.

I stared at the cold, gray rapids thundering through the cold, gray fog. Why was I here, alone, in the middle of nowhere? Why did I keep going out on my own into the wind and the rain and the wet? I inhaled the cold, moist air and searched the shadows beneath the Khembalung caves, searched the thick fir trees clinging to the edge of the valley floor. The mist moved swiftly and silently along the banks of the Barun. The silty river roared. Then the clouds suddenly opened, and a shaft of light broke through the fog, turning the gray water to silver, the black fir to deep green. An angular cliff appeared out of the clouds high overhead; the red-gray granite, softened by the yellow evening light, was framed by heavy dark clouds. Just as suddenly, the fog closed over and night set in.

I took a deep breath, turned, and walked back to help prepare dinner. In the cave, a young woman approached me to say that, unlike me, the other pilgrims were with their families and neighbors, that it was as though they had never left home. They forgot that I was alone, and she realized that sometimes I must feel lonely or homesick and hoped that I was okay.

During the trip, I felt an ache in my chest, a longing that would not go away. I thought there must be a place, somewhere, where I could be held, here, on the inside. As soon as I arrived at that place, I was sure, the yearning would disappear. Now I realized that the longing would never go away. It was what had brought me here, to one of the thinnest places I had ever encountered. It was what let me know that I was on the right path, what kept me walking even as the skin peeled off my feet. It was what let me experience the sacred.

◊

Several months earlier, I had been sitting on the bank of the milky green Arun River to witness a cremation. As we watched the burned body float down the river, the mother of the dead man held up her hand in front of my face. It was cracked and dark. "We all feel love," she told me. "We all feel pain. We all bleed when we are cut. It is only the *mindhum* [oral traditions] that is different."

The skin contains the blood, creating the distinctions that enable us to live. But it can become too thick, letting us forget the blood pulsing underneath and keeping us from sensing what Roberto Calasso calls the "connection of everything with everything else, which alone gives meaning to life" (1993:284).

We experience the sacred not simply by visiting places that are sacred. We enter the sacred when we let go of the fear of being exposed and begin to open our hearts to the world around us. Only when I gave up trying to hide what was inside did the boundaries between the other pilgrims and me begin to dissolve. And in the moments I felt most alone, I realized that I was never alone. "Longing begets belonging," writes Mary Oak (2000). The sacred, as Calasso writes, is always "waiting to wake us and be seen by us, like a tree waiting to greet our newly opened eyes" (1993:280). It is simply up to us to let ourselves see.

◊

Having reached our destination, everyone was suddenly in a hurry to begin the trip home. Rice fields had to be weeded, millet had to be planted, and houses had to be looked after. We left early the following morning; walked for twelve hours, over a 5,000-foot scree pass; and then descended steeply past grazing yaks and shepherds' huts. We walked on after dark for an hour, searching for a place to spend the night. Finally, ten of us crowded into a small, empty bamboo hut. I was the only one with a sleeping mat, so I kicked away the cow and goat dung, spread out the mat on the dank floor of the attached livestock shelter, and tried to sleep. We again awoke before dawn and walked hard and fast until we reached another shepherd's hut where we stopped to drink sour buttermilk. Since climbing to the caves, I had given up bandaging or even looking at my feet, but by this time, I was no longer the only one limping. The women leaned heavily on walking sticks and groaned with each step. We joked and laughed to keep our minds off the pain.

The trail descended steeply. Yaks gave way to water buffalo and cows, and we began to meet shepherds from Hedangna. Finally, we could see the village, far down the ridge. We had been rushing and rushing, and now the women wanted to linger, to hold onto the remaining bits of time that were outside regular, routine time. We paused on top of the ridge to eat the last of our corn flour. One woman sighed and said that she was so happy up here, in the meadows and the mountains, that she did not want to return home. Two women separated the tiny wildflowers they had collected from the fields beyond Yangle Meadow to give to friends who had had to stay home. Two others divided a bottle of water, taken from the headwaters of the Barun. The sun was beginning to set.

We began to walk the last stretch, down and down. We came across leeches for the first time but were too tired to pull them off. An hour later, we entered the outskirts of the village, in the dark. I was the only one with a flashlight, but the batteries were weak, so our pace slowed to a crawl in the dim glow. The trail wound beneath thick clumps of bamboo towering over the stone and mud houses. People broke off from the group as we passed the narrow paths to their homes. Eventually, it was only the two grandmothers and me, walking down to the houses at the bottom of the village. We finally arrived. I dropped my pack and leaned it against the stone wall of Ganesh and Jaisita's home. Someone went inside to cook rice. The children gathered around while I sat on the mud porch to unlace my boots. My socks were wet with blood. I carefully peeled them off, so the air could begin the slow process of healing—and thickening—the exposed raw skin.

◊

Only much later did I realize that this pilgrimage had marked a shift in my journey. I had crossed a threshold, the crumbs marking my way home had been eaten. The only way now was to go deeper into the forest.

11

THE SACRED SPRING

October 1992

O ne afternoon in early autumn, after the pilgrimage to Khembalung, I stopped by Baiseti Thuma's home. She was on her porch, holding a leaf cup filled with *jad*, when I walked up. She told me that she was going to the millet fields by her house to leave offerings for Chaketangma, the wrinkled old ancestor who was said to wander through corn and millet fields and who, I imagined, looked a lot like Baiseti Thuma. Chaketangma had been making her eyes itch, Baiseti Thuma explained. She had forgotten all week to make an offering, and today she was finally going to do it.

She started to walk away, and then paused and turned back to me. She said that she had overheard me on my porch, transcribing an interview, with the help of Raj Kumar, about land surveys and disputes. "All that about *kipat* and headman that Raj Kumar was talking about, that's all *bharkarko kura* [just now information]," she said. "If you want to know what matters, you need to learn about the *mindhum,* about the *mindhum* and the *tsawa.*" And with that, she turned, shoulders bent, and headed off to find Chaketangma.

Emboldened by the pilgrimage and admonished by Baiseti's words, I decided again to approach the priests and shamans. This time, as she instructed, I started by asking about the *tsawa.*

◊

The *tsawa* is a pool of clear water drained by three stone spigots on the northern edge of the lower village of Hedangna. Huge pipal trees, which people are forbidden to cut, tower over the platform, making it a cool, peaceful retreat even on hot, dusty days in May. This pool is the remnant of the deep-blue lake from which Minaba and Sepa drank after having sworn to stay on these lands. The spring from which the lake flowed thus became the brothers' *tsawa.* Years passed and the lake dried and all that remains is this water hole, with five tall stones placed in the center when five ancestors celebrated the founding of the village. They named the spring Hanghong *tsawa.* "Hanghong" means "to celebrate, to make offerings." Villagers say that several stalks of rice grow by the pillars each year without ever being replanted, and they whisper that on some days, those who listen closely can hear the ancestors laughing, singing, and clapping their hands.

Each April, before the preparation of rice fields began in earnest, men and some women from the upper and lower villages gathered at Hanghong *tsawa.* They brought containers of *jad,* small sacks of rice, and some money to contribute to the cost of the chickens. Some carried loads of firewood and large metal pots. The first to arrive got right to work, building fires and filling pots to cook the rice. One or two climbed down to the woods below the stone platform to gather branches and bamboo to prepare the altar. Those who arrived later sat cross-legged on the large flat stones, sipping a *tongba* (another way of drinking millet beer) and watching the preparation.

One April, Purnamba, considered the most powerful priest in the village, arrived just as his assistant finished building the altar. He wore a short white tunic tied loosely at his breast. His bare legs were sculpted by years of hard work on steep, rocky land, the calves slightly bowed. His hair was white. He made a few final adjustments to the altar, picked up a brass urn filled with wildflowers gathered from the banks of the stream and the edges of the fields, and stood before the five stones.

He closed his eyes and began to chant the name of the *tsawa,* Hanghong *tsawa,* over and over. He recited the tale of Yamphuhang, the Yamphu's ancestor, making his way up the Arun Valley, looking for a home. He told of Yamphuhang's stay in Tibet and Minaba and Sepa's journey to find a new home. He described the five original clan fathers and the first ceremony at Hanghong *tsawa*, telling how the ancestors placed these five stones in the pool of water to mark the founding of the village.

Purnamba shook his bamboo wand and the urn of flowers. His voice rose and then fell, and his body began to shake. One moment he was there, before the five stones; the next, he was walking up the hill, through the fields, and into a forest. The forest was dense and lush, the trees much bigger than they are today. He walked slowly, aware of each foot as he placed it on the forest floor—watching, watching, walking, walking—uncertain of what he would find.

He chanted the name of each flower and bush that he saw, called by name each bird that flew overhead. Water rushed over the rocks of a stream. The wind blew against his cheek. He passed through shadows and into pockets of sunlight. He kept walking, kept looking, kept listening, kept waiting. From deep in the forest, behind the dark trunks, a shape emerged and then another and another—until there were five in all. Purnamba raised the urn of wildflowers and bowed his head, grateful that his chanting was working. He shook the urn and repeated his greeting, "*Sha'de! Sha'de!* [I greet the God within you!]."

The priest was still standing on the stone platform before the five stones. The men seated on the platform saw only his body; they heard only his words. They could not hear the birdsong or the rush of the water, could not feel the wind or the crunch of leaves underfoot, could not smell the flowers, and could not see the trees or the five founding fathers of the village emerging from the dark woods. They did not even know that the ancestors were before them until Purnamba called out their names one by one, greeting them and inviting them to join him, to come back to the *tsawa* to partake in a feast. As he waited for the shades to make their way through the forest, he chanted the story of how these shades had taught the Yamphu about *charawa*, a Yamphu concept that refers to the essence of grain, and how they brought that same essence now:

Dancing, laughing, playing and singing they ask the ancestors for charawa.
They ask for charawa *for the rice that grows in the place where the sun rises;*
They ask for charawa *for the rice that grows in the place where the sun sets;*
They ask for charawa *for Mades rice, for Bhotiya rice, for the rice of*
  Hedangna.

When they arrived at the *tsawa*, Purnamba urged the shades to eat, drink, and be merry. They laughed and sang and clapped their hands. After eating the chicken and the rice, after drinking the *tongba*, the priest chanted, it was time to continue their journey.

The chant continued, describing how the ancestors loaded the *charawa* into bamboo baskets, lifted the baskets onto their backs, and began to carry the heavy loads up the steep slope. They stopped to rest on other stone platforms in the cool shade of pipal trees, and the priest sighed to express their fatigue. They laughed and sang and danced as they went. They took their time. They drank beer. When they reached the home of the assistant village headman, the ancestors called to the woman of the house not to be angry; they called to her to come and get this load of *charawa* and to fill her storage bin until it overflowed.

The *charawa* delivered safely into the storage bin and the ancestors' appetites satiated, the priest recounted how he led them back down the hill, through the *tsawa*, and up the hill on the other side, chanting the names of the trees and the flowers to mark their progress for those left behind. Once they arrived at the edge of the forest, he bid the shades farewell. He raised the brass urn filled with flowers, bowed his head, and repeated, "Sha'de! Sha'de!" He turned; walked back down the hill, more rapidly this time; and passed quickly through the *tsawa*.

Purnamba stopped chanting. He knelt to place the brass urn on the stones before the altar. He told his assistant to gather the dead chickens and the plates of rice, and he joined the men by the fire.

There is one *tsawa* for the village and other *tsawas* for each clan. Each *tsawa* is sacred, particularly Hanghong *tsawa*; they are kept clean, and the surrounding trees are protected. The *tsawa* is also something inside. It is like an address, villagers told me. They use it to identify themselves to strangers, to one another, and to the ancestors. When they meet someone for the first time, they introduce themselves with first their clan and then their *tsawa*, thus claiming their place in the Yamphu social world. Politically, the *tsawa* proves that they are either the original or the adopted descendants of the Yamphu ancestors and thus gives them the right to claim land as *kipat*, extending a cultural marker into the political landscape. The *tsawa* is also the doorway through which the shamans and priests pass on their journeys into the realm of the ancestors; it is literally the place where humans can contact the divine.

When performing the *sammang*, the most important ritual for individual health and healing, the village priest identifies the individual to the ancestors by calling out his or her *tsawa*. In the same way that membership in a clan places the villagers in the social world, the *tsawa* places the members of the clan in the spirit world; it is a name, in a way, for their collective soul, the spiritual essence of their community, an essence that literally is the spring from which their ancestors first drank. It is a kind of spiritual home, a centering force inside made stronger through its proximity to the actual site of the *tsawa*. Its location is important. Men and women speculated that the men who lived in the lower village did not drink as much *jad* or play cards as frequently as the men who lived in the upper village because they were closer to the *tsawa*, because they literally drank from that water each day. Drinking that essence, the villagers seemed to be saying, had a moderating effect on the men's behavior in the world.

◊

Tsawa and *charawa* were the most difficult Yamphu concepts to grasp. There are no equivalent terms in English. Despite or because of their untranslatability, I felt that they were the most important ideas for understanding the Yamphu in general and, more specifically, for understanding their relationship with the land and their concept of home. The point of the *tsawa* ritual is to imbue the grain with *charawa*. Just as the *tsawa* identifies the spiritual essence of an individual, I understand the *charawa* to be the sacred essence of the grain. This spirit enlivens the rice, making it last much longer than the material substance alone would ever last.

Although the villagers are utterly absorbed in the physicality of their lives—the sun and soil, the wind and rain—the nonmaterial world permeates the material world. When they eat the rice harvested from their fields, they eat from the hands of the ancestors. When they drink from the *tsawa*, they drink the waters that quenched the thirst of their forefathers. Spirit and matter merge in the *tsawa* and in *charawa* in a way that is hard for my mind, raised in a culture that insists on their separation, to grasp or to articulate. As the literary historian Robert Pogue Harrison writes, "The originating source itself remains unspeakable. . . . It lies behind the landscape, to be sure, yet not like a face that hides behind a mask. It is nothing other than the landscape in its unaccountable presence. As it withdraws behind the appearances, it leaves in its place a landscape" (1992:241).

The goodwill of the ancestors imbues crops with an essence, *charawa*, that nourishes and sustains the community in the present. The glacial lake in the Poptil La, on the border between Tibet and Nepal and a three-day walk north of Hedangna, described in the founding myth, is believed to be the source for Hanghong *tsawa*. When the water in this lake is low, Yamphu older men and village healers say, the productivity of their rice harvest will be poor (it will have no *charawa*). When the water is high, their rice harvest will be good (it will have *charawa*).

For the Yamphu, the *tsawa* is an opening, like the Khembalung caves, where the longing of the land—the voice of the ancestors—can be most easily heard. It is a thin place between what poet David Whyte describes as the world of dance and inclusion and that of exile and aloneness, the place where the longing of a people and the longing of a place intersect.

◊

Years later—after returning to the United States; moving from Cambridge, Massachusetts, to Canaan, New Hampshire; and giving birth to a daughter, Avery—I went for a hike up a nearby mountain one autumn day. Avery was six months old. It was cold and late in the day, but it was a short climb to the top. I thought that if I walked quickly, I could make it back to the car before dark. Avery soon fell asleep to the rhythm of my walking and remained so until I reached the top of the mountain. Just as I turned to hike back down, the wind woke her and she began to cry. I walked for a while, hoping that she would settle back down, but she kept crying. I finally stopped to see if she was hungry. She nursed a bit and again began to cry. The sun was beginning to set; I put her back in the carrier on my chest and walked more quickly beneath the darkening sky. Her cries got louder and more persistent. I stopped again, but that just made her angrier, so I carried her in my arms the rest of the way down, singing songs, making sounds, doing whatever I could to calm her. She was shrieking by the time we reached the car, and she kept it up all the way home, not stopping until we were inside and she was out of her jacket and back in my arms, nursing herself to sleep.

This episode was not a big deal. Avery was fine; I was fine. And yet I ached inside, a sense of some ancient kind of betrayal that was not rational. I was not sad because I had failed to anticipate her needs or had caused her to suffer by fulfilling my desire. That was part of motherhood. It was bigger than that. While Avery was in my womb, her needs were met. She was always warm; food was always available; she could sleep and move as she wished. My body—the landscape from which she was born—had contained her, entirely. And now it did not.

A few nights later, I lay on our bed, Avery at my breast, nursing herself to sleep. I listened to the rain on the roof and thought about what it meant to drink water from the *tsawa*, whether the Yamphu considered it to be different from drinking water from any other spring. I thought about what it meant to eat food embued with *charawa*, with the essence from the ancestors that made it last much longer than the physical grain would otherwise last. I thought of Avery nursing, of what she received along with the milk, and of bringing her to my breast when she was agitated and how, by finding her center in the land-scape of my body, she remembered that center in her own.

The *tsawa* is a place in the landscape where the Yamphu can feel close to the ancestors in the way I imagined Avery felt close to me when she was nurs-ing in my arms. No distance between us—remembering what it had been like to be one. The Yamphu cannot stay at their sacred spring forever, drinking the waters of the ancestors. They have to return, alone, to their fields and homes. They have to live their own lives in their own ways. But the *tsawa* is always there, a place in the landscape to which they can return, again and again, to experience a kind of wholeness, a place in the present they do not have to go back in time to recover.

I interpret the presence of this spring in their lives, a centering force they have never not known, through the absence of such a center in my own life. If, at any time, we drink enough from that spring, from any spring that is sa-cred—I think of Avery at my breast—will it fill us for a lifetime? Are our lives then shaped by a sense of fullness at our core, not one of emptiness?

When I was in Hedangna, I thought that the Yamphu could hear the voice of the sacred so much more clearly than I because that sacred otherness was present in their landscape and culture in ways it no longer was in my own. After leaving, I began to wonder if I had it wrong. I began to think that perhaps it is not that our land no longer speaks, but that we do not have a concept—a *tsawa* or *charawa*—to alert us to the possibility that the land can speak, to re-mind us that the ancestors are still present. We have not been taught how to listen or what to listen for. It is not that the sacred is absent, but that we no longer know it when we see it. And we have no guides—or do not recognize them when they appear—to teach us how to respond when we do.

12

KELEKPA THE SHAMAN

October 1992

Once, so the stories go, there was no division between humans and the ancestors or between those who could speak with the ancestors and those who could not. At that time, the ancestors did everything humans could do, I was told, but they did it better, longer, more easily. Humans had to climb steep mountains, but the ancestors could fly over the land like birds, soaring from the summit of Makalu all the way to Hedangna in the time it took to finish one long whistle. Humans aged, grew weak, and eventually died. The ancestors never died. Those who were old had always been old, and those who were young stayed young forever.

Even with these differences, humans could see the ancestors just as clearly as the ancestors can now see humans. They could walk together and eat the same food. They could even fall in love with each other, as did a young Yamphu man and the daughter of the most powerful and dangerous of the ancestors:

Matlung Thuba. Even today, villagers in Hedangna speak Matlung Thuba's name only in a whisper, glancing cautiously over their shoulders in fear that, if the ancestor heard his name, he would suddenly soar down to the village and snatch someone's soul. This young man was brave or foolhardy enough to ask Matlung Thuba for permission to marry his daughter. Although such an alliance was unusual, the youth seemed strong, responsible, and honest. So after much pleading from his daughter, Matlung Thuba agreed. The two married. They moved into the same house and they farmed the same fields. It was as though there were no differences between them at all.

After several months, the young man (called *maksa* [son-in-law]) joined his father-in-law and brothers-in-law on a hunt, setting out at dawn to travel through the dense woods along the banks of the Arun River. They walked and walked. After several hours, they turned up the Barun Valley, which is even narrower and steeper than the Arun. As they walked, the brothers talked about the *ledey*, a kind of prey, that they were hunting. The *maksa* had no idea what a *ledey* was, but he was eager to please his in-laws and so continued to look for an animal that might be this *ledey*.

Finally, late in the day, a tiny bird darted down the valley. The brothers cried out, "A *ledey!*" They immediately began to chase what to the *maksa* seemed like invisible prey down the Barun Valley; along the Shingsa ridge; past Terathum, a week's walk to the east for humans today; and then to the Bakre River, far to the south. Up and down they ran, unable to trap the *ledey*. Finally, Matlung Thuba told his exhausted son-in-law to stand at the end of a small valley. The others would chase the *ledey* up the valley, where the *maksa* was to shoot it with an arrow.

The *maksa* climbed to the end of the valley and sat down to wait. He still had no idea what a *ledey* was. For him, "prey" meant a deer or a bear. He had seen no sign of either all day. He sat and watched the shadows cast by the evening light, thinking about how tired he was and what a wasted day it had been, longing to be home by the fire and drinking *jad*.

He noticed a tiny bird flying up the valley. Bored and not thinking much about what he was doing, the *maksa* shot it with an arrow. He looked at the bird briefly before stuffing it under his hat and promptly forgetting about it. He sat back down to wait. After some time, his relatives climbed up the valley and asked whether he had killed the prey they had sent his way. "Prey? No prey has come up this valley," he said, shaking his head. Perplexed, the in-laws went back down the valley to look for the *ledey*. Two or three more times, they returned to ask whether the *maksa* had caught the prey. Each time, he slowly shook his head and said that no, nothing had come his way. After the fourth round of questioning, the brothers looked so confused that the *maksa* reached under his hat and pulled out the small bird. "This is the only thing that has come up the valley since I've been here," he said. His in-laws raised their arms in dismay. Matlung Thuba cried, "But this is what we have been chasing all along! What did you think we were looking for? What a useless son-in-law!"

Clearly disgusted by the stupidity of this human, the brothers began to clean the bird. One collected leaves in which to carry the meat home, another washed out the intestines, while another carefully divided the bits of meat into tiny piles on the leaves. The *maksa* had killed the bird, so he was given the thigh. The bird was so small that its entire body could fit in the palm of a hand, and the thigh was smaller than the *maksa*'s tiniest finger. He watched with astonishment as each brother carefully wrapped his even smaller portion of meat in large leaves, placed it inside his shirt, and happily set off down the trail.

The *maksa* followed, thinking "What in-laws I have. They are Kiranti, and yet they don't even know how to hunt! All this time we have taken, all this running around, and we haven't even gotten enough meat to feed ourselves one meal." They arrived home long after dark. He sullenly said goodbye to his relatives and strode into his house, where his wife was tending the fire.

"La!" he cried out, throwing the thigh at his wife. "This is what your family hunts for! Here is the meat we are meant to live on!" As he threw the piece of meat, the tiny bone suddenly became quite big. It landed with a dull thud on his wife's leg. She shrieked in pain. "What have you done?" she cried. Her father rushed in to see what had happened, and she turned her anger toward him. "Why did you do this to me?" she cried. "Why did you marry me to this big meat-eating, buffalo-eating human! This bone is plenty for us, but it has no meaning for him! Look—he has broken my leg! What am I to do?"

The storytellers paused at this point to underscore the blindness of humans. Although tiny, the bird would provide food for the ancestors for a long, long time, they explained. The problem was that the *maksa*, a greedy human, could see only what was visible to his eyes; he believed only what he could touch with his hands. He did not understand that a body needs more than physical sustenance to stay alive and did not trust that the nourishment offered by the bird was as sustaining as that provided by a larger animal, such as a bear. Size has nothing to do with it.

◊

Shocked at how narrowly humans saw the world and at how selfish this myopia made them, the ancestors decided to move farther up the ridge; from now on. they would live separately from and be invisible to humans. Although humans could no longer see the ancestors, the ancestors could still see humans. They could enter their homes and eat their food, slide into their dreams and snatch their souls. And as they disappeared from view, the ancestors designated a special class of villagers who, under certain conditions, could still see them, could hear their requests, and would know how to meet their demands.

Unlike Hindu priests and Buddhist lamas, these healers—the shamans and priests of Hedangna—do not undergo any formal training. There are no written texts to describe the types of visions they might have, no institutions to sustain their work. Shamans and priests are distinguished, simply, by their dreams. While everyone else hauls firewood and cuts rice in their dreams, shamans swim in a cold clear lake on the eastern ridge of Khembalung. What they learn on these nocturnal journeys enables them to "see double"—to see the visible and the invisible—and so pass through the "ethereal veil" wafting between the living and the nonliving and enter the realm of the ancestors.

◊

It was autumn. I had been in Hedangna for ten months. Brian had recently returned to Nepal and, after a month trekking, had come to spend two and a half months in the village. We were sitting with Devimaya in the sun, waiting until the rice was cooked. Kumari suddenly ran out of the house, carrying a large tin bowl of freshly ground millet flour to show us a design written lightly on the surface of the flour. The bowl had been left out overnight, and in the morning, when Dhanmaya had discovered the markings, she asked her children who had written them. None of them had.

Devimaya said this meant that the pattern had been made by a *shirawa*, the wandering soul of a living person that had been snatched by the ancestors. Kumari then mentioned that several weeks earlier, the family had come downstairs in the morning to find dishes spread across the floor. None of them could have done it, she said, since they were asleep. If things like this happened outside the house, it meant that someone in the village, a relative or neighbor, would die, Devimaya said. When they happened inside, it meant that someone in the house would.

Devimaya said that she felt like crying. Kumari was afraid to go into the house. I translated their conversation to Brian. He then told me about a dream he had had a few nights earlier. Kumari was being carried to Khandbari on a stretcher. Brian was walking beside her, and she was dying.

I did not translate this dream to Kumari and Devimaya. Kumari had been sick off and on since I had moved in with her family. A shaman once told her parents that she would either die young or live a very long time. I told myself that shamans rationally could not make such predictions and that a mouse had made the markings in the flour. Who knew about the dishes on the floor, but I was sure that there was also an explanation for that. What the sisters said made me uneasy, though. In spite of my skepticism about superstitious beliefs, I still feel funny when I walk under ladders and always knock on wood when I talk about something bad happening. Brian's dream just made it worse.

The following week, Dilli Prasad and Dhanmaya called Kelekpa, the shaman, to their home to discover why Kumari had been sick for so long. A crowd gathered on the porch to watch the ceremony. Shadows from butter lamps flickered across the dirty, round cheeks of children crowded on straw mats, their bodies almost invisible in the moonless night. The coals from the fire had died down, and the children, dressed in thin cotton, huddled together under woolen blankets to keep out the cold. Some slept, others fidgeted as they tried to get comfortable, but most stared intently at the men who were cutting branches and helping Kelekpa's assistant prepare the altar.

The assistant laid broad-leafed branches against the whitewashed wall, creating what looked like a forest garden on the edge of the mud porch. He stuck tridents and mountain-goat horns among the leaves. Shiny brass plates placed on the ground in front of the branches were piled high with rice and a 50-rupee note, Kelekpa's pay for his evening's work. A *kukuri* (long, curved knife) lay against the branches, alongside a bamboo container filled with millet, another *tongba* for the shaman after he completed his work. The air was filled with the scent of burning juniper, collected by Dhanmaya in August on the pilgrimage to Khembalung.

Each time I saw one of these altars, I pictured a peaceful, quiet clearing in an evergreen forest, pine needles underfoot, clear and open, an enchanted and magical place where spirits—and lost souls—must want to linger. That, Devimaya told me, was the point.

Kelekpa arrived just as his assistant completed the preparations on the porch. Kelekpa was taller than most Yamphu men, thin, with thick dark hair that fell across his eyes. He drank too much and, when he was drunk, fought, especially with his wife. But he was also a storyteller and a performer. He had a kind of charisma that other men in the village lacked. He danced a lot, Devimaya said, and he was the only Kiranti shaman in the upper village. Most important, he usually found the souls he was sent to recover.

Kelekpa unwrapped several bundles of cloth taken from his shoulder bag and took out two large chunks of quartz that he placed at the base of the branches. He hung a betel-nut necklace around his neck, tied a string of bells around his waist, wrapped a turban around his head, and stuck a peacock feather in the side. He did each action carefully and precisely. He was about to travel into a different order of reality. He could not be careless. He took a deep drink from the *tongba*; picked up his wand, a long piece of bamboo with frills shaved down at either end; and began to chant his *mindhum*, Yamphu stories of the creation of the world.

By chanting the *mindhum*, Kelekpa contacted his guru, who, in turn, enabled him to "see double" and enter the realm of the ancestors. Kelekpa's chant became louder, the clang of the stick on the bronze plate became more rapid, and his body began to shake. He called his guru by name. The children watched intently, absorbed by the drumming on the plate and the chanting, but others on the porch continued to talk and tend the fire. Kelekpa did not need their help; he would let them know what they needed to know when the time came.

His voice became louder. His body began to shake more forcefully. Suddenly, he began to dance, back and forth, across the porch, his motions still intense but smoother than before. He kept dancing, sweating from the heat of the fire; suddenly, as he entered his trance, the smoke and the voices became distant and muffled. His guru had arrived, lifted his soul from his body, and carried it down the hill to the *tsawa*—the sacred spring of the ancestors.

At the *tsawa*, he bowed, greeted the ancestors—"Grandfathers, Grandmothers, "*Sha'de! Sha'de!*"—and continued on his way, through the *tsawa* and to the south, riding the currents of the wind. He chanted the tributaries of the Arun River—Tamor Kosi, Dudh Kosi, Tama Kosi, Sun Kosi—marking his progress for those left behind. He continued southward, following the course of the Arun and looking for signs of Kumari's *lawa* (soul) in narrow valleys and along desolate ridges. Seeing nothing, he passed the Nepalese border and soared over the flat, hot plains of India. In a matter of minutes, he arrived at the place called Kasi, the entrance to a dark tunnel that drops eleven levels into the center of the earth and through which only priests, shamans, and witches—those visited by the ancestors in their dreams—can pass.

Two fierce lions blocked the entrance, gatekeepers to ensure that only those who are ready can enter the worlds below. Kelekpa raised his bamboo wand, bowed his head, and chanted "*Namaste*" to the gatekeepers, "*Namaste*" to the ancestors: "*Sha'de! Sha'de!*" He waved his wand and chanted, again and again. The beasts finally stepped aside, and he continued on his way.

Kelekpa had been traveling through open space, flying high above the land, but now he entered the narrow tunnel. A chill wind came up from the darkness, pressing his tunic against his chest. It was too dark for shadows. He was absolutely alone. He chanted with even more concentration and intensity than before, "*Sha'de! Sha'de!*"

He reached the next gate. More beasts loomed on either side, blocking the path. They demanded to know his business and whence he had come, hoping to frighten or tempt him from his journey. He bowed his head and raised his wand: "*Sha'de, Sha'de!*" He eventually was allowed to pass. Back into darkness, another gate, more chanting, an opening, and again into darkness. He held his ground before the gatekeepers at each entrance, bringing his attention back, again and again, to his chanting, not giving in to the silence or the darkness, not letting the absoluteness of his aloneness undermine his intent. The more he was able to contain his fears of failure and death—the more he was able to hand himself over, utterly, to the power of his guru carrying him along—the more likely he was to reach his destination. He chanted and chanted until, finally, he was once again swept through the doors and deeper into the earth.

Kelekpa finally reached the *hiwalok tem,* the place of the wind. A ladder led down from the narrow tunnel into a cathedral-like cavern. A large, clear lake emerged from the shadows. At the far side of the lake, on a stone platform under a pipal tree, sat an old, old man. His hair was white, and his face was creased with the wrinkles of a thousand years. He was turned away from the shaman, looking silently into the distance. This man was Manguhang. He was older than everyone; he was here at the beginning of time. He was the source of all that was good and all that was bad in the Yamphu world.

Kelekpa descended the ladder to the edge of the lake, shaking his bamboo wands and repeating "*Sha'de! Sha'de!*" Witches, priests, and shamans stood in three separate lines. He joined the line of the shamans. When his turn came, Kelekpa stepped forward and stated why he had come. He continued to chant, head bowed, as he awaited Manguhang's reply. When he received an answer, he thanked Manguhang, turned, and rode the wind out of the darkness, through the *tsawa,* and back to Dilli's porch.

Kelekpa's body suddenly stopped shaking. His right hand was clenched, and he thrust it toward a bronze bowl filled with water. He peered into the bowl to see Kumari's soul: a long, thin white soul meant health; a short black one meant death. Kumari's was white and thin. Kelekpa cupped his hand over the dish and came to Kumari's side. He blew on the crown of her head and into each ear and waved his wand over her entire body. The soul had been recovered. The relief among Kumari's family was palpable, their fears about Kumari erased, at least for the time being.

Kelekpa returned to the altar, chanted a bit more, and then stopped. The drumming stopped. Somewhere down the hill a dog barked. Kelekpa adjusted his headband. He knelt and leaned to the bamboo straw to take a long sip of his *tongba*. He sat back on a straw mat and began to describe where he had been.

We all had heard stories of the realm of the ancestors passed on by the shamans and priests who had traveled through that domain, but no one in the room had made the same journey. Only a few could understand the actual meaning of Kelekpa's words, which were spoken in a ritual language reserved for those who learned it in their dreams. All we could see was Kelekpa's trembling body. All we could hear was his rapid chant. All we could know of this particular journey was what he would tell us, later, after his return. Kelekpa could cross a boundary that the rest of us could not.

Kelekpa's journey is the hero's journey. He is led by a spirit guide across a threshold and plunged into darkness, where he encounters horrors too great to imagine. With the help of his guru, he holds his course until he reaches sacred ground at the center of the world. There he receives a blessing that he carries back to the world of the living, where he uses the knowledge he has gained to heal.

◊

A few weeks after his performance, Kelekpa happened to be drinking *jad* at a house I visited. The weather prevented him from working in his fields, so he volunteered to let me record the ritual chants, the *mindhum*, needed to make his journeys. Surprised but not questioning his willingness to do now what he had refused before, I turned on my tape recorder. He held the microphone close to his mouth and began to chant. As he chanted, he began to sweat. His breathing became difficult. Stopping to catch his breath, he mumbled that he should not be doing this. At one point, he asked the owner of the house for some rice to give as an offering to Matlung Thuba, his main guru. Matlung Thuba lived far up the ridge, toward Makalu. When Matlung Thuba stole a soul, the person was likely to die. Kelekpa explained that once he had gotten Matlung Thuba's attention, the ancestor would expect a goat or a chicken. If he did not receive one, he would become angry and refuse to come when Kelekpa called him in the future.

When he finally finished, Kelekpa said that he had wanted to teach me everything about his *mindhum*, the origins of all the ancestors. But, it seemed, his guru would not allow him to do so. Raj Kumar seemed skeptical, but he seemed skeptical of most things that had to do with priests and shamans in the village. Harilal, a gaunt fifty-year-old man sitting next to the shaman, looked at Raj Kumar and said solemnly, "You don't believe this, but we do. For us, it is real." Later, Kelekpa told Raj Kumar that he had been sick for two days after chanting the *mindhum* into my tape recorder.

Priests and shamans were secretive about their work in part to protect their monopoly as village healers. But it was more than that. The ways that Kelekpa and Purnamba described their work to me underscored its gravity. Communicating with the ancestors and attempting to interpret their needs and desires was a deeply dangerous undertaking, to be attempted only by those with the proper training. The words that Kelekpa chanted into my tape recorder were powerful. They were words that he used to travel between worlds; they were words that could make spirits move and were not to be treated lightly.

Although I was surprised at the effort that Kelekpa and, later, Purnamba made to help me understand, particularly after their initial reluctance, Raj Kumar was not. In the beginning, he said, people did not know who I was. I was from the outside and not to be trusted. Now they knew me, and they wanted to help. Helping was a way to show intimacy, he said. I think that the healers were also more willing to speak to me after I had "proved" myself on the pilgrimage to the Khembalung caves. Both Kelekpa and Purnamba often referred to the pilgrimage during our conversations. And once I became Devimaya's *mitini* (ritual sister), both Purnamba and Kelekpa became my "relatives." Finally, I think that both of them were intrigued to know that their chants would eventually reach and be heard in the United States. In any case, their willingness to reveal what had initially been "off-limits" said something about the permeability of the boundaries between priests and shamans and everyone else. Like the boundaries of *kipat*, the relationship with the person on the other side was ultimately what mattered.

An anthropologist at a Nepalese studies conference in Kathmandu once pointed to another anthropologist and said with disdain that that person actually believed that what the shamans did was real. I did not respond. I did not know if the shamans could actually communicate with the ancestors or if they really did see lost souls. It was real to them, and that was enough. Even Raj Kumar, one of the most "modern" Yamphu villagers, felt uneasy when rituals were not performed correctly, and he was surprised when another Yamphu schoolteacher in the village got rid of his household shrine and refused to offer new rice to the ancestors. I was interested in what the priests and shamans did because I was interested in what mattered to the Yamphu. And I was interested because I wanted to know what it took to experience the sacred at home and not just on a pilgrimage to a far-off place.

◊

In the late autumn, as my first year in Hedangna came to a close, I read William Faulkner's short story "The Bear" for the first time. I had not read much by Faulkner, but just before leaving for Nepal I had read an essay about "The Bear" by Donald Snow (1991). Snow argues that because the conservation movement has taken "God out of the loop"—because it is not based on an intrinsic sense of the sacredness of the earth—it can never be the solution that conservationists hoped. Snow retells the story of "The Bear" to illustrate what he means by this sacredness, how we should act in its presence, and what happens when we do not.

As I sat on a stone platform overlooking the rice fields reading "The Bear," I was struck by how Faulkner captured what I never could about the changes in the land that I was observing in Hedangna. He got inside the story and lives of his characters in ways I could not—or did not dare—as an anthropologist intent on documenting what I observed in the world around me, not what I imagined.

The power of Faulkner's writing emerges less through any isolated passage than in the way the story circles and builds, adding layers to what started as a straightforward tale about hunting a bear in the wilderness and ultimately becoming an indictment of the worldview that assumes that people and the earth can be owned:

> Bought nothing. Because He told in the Book how He created the earth, made it and looked at it and said it was all right, and then He made man. He made the earth first and peopled it with dumb creatures, and then He created man to be His overseer on the earth and to hold suzerainty over the earth and the animals on it in His name, not to hold for himself and his descendants inviolable title forever, generation after generation, to the oblongs and squares of the earth, but to

hold the earth mutual and intact in the communal anonymity of brotherhood. (1961:257)

For me, the heart of the story comes when Ike, a young boy and the main character, set out to find Old Ben, the great marauding bear that for years had eluded the hunters' bullets in Big Bottom, the vast wilderness in Mississippi where Ike has gone hunting each year with a group of older men. Eventually, Old Ben would be shot. Ike knew that. Before it happened, he wanted to see the bear, not to kill him, but simply to be in his presence. And so during one autumn hunting trip, Ike left camp each morning before dawn to spend the whole day in the woods, searching for signs of the bear. The men thought that he was hunting squirrels. Only Sam Feathers, "half-Chickasaw, half-black, more like Ike's father than his real father," knew what Ike was doing. On the third afternoon when Ike returned to camp at twilight, Sam said, "You ain't looked right yet." And then he said, "It's the gun. . . . You will have to choose" (1961:206).

The next day, the boy set out again before dawn, this time without his gun. He hiked until noon, still with no luck, and suddenly realized that leaving the gun was not enough. And so he took off his watch and his compass, hung them on a branch, and kept going. When Ike realized that he was lost in the wilderness, nine hours from camp, instead of panicking, he did as Sam Feathers had "coached and drilled him"; he circled slowly, then circled again. Still he did not find the tree where he had left his compass and watch. "And he did what Sam had coached and drilled him as the next and the last, seeing as he sat down on the log the crooked print" (1961:208): the track of the bear. He saw another. And another. He followed the tracks until he emerged into a glade:

It rushed, soundless, and solidified—the tree, the bush, the compass and the watch glinting where a ray of sunlight touched them. Then he saw the bear. It did not emerge, appear: it was just there . . . then it moved. It crossed the glade without haste, walking for an instant into the sun's full glare and out of it, and stopped again and looked back at him across one shoulder. Then it was gone. (209)

◊

Unlike Tibetan lamas, who meditate to tame the wilderness inside, priests and shamans in Hedangna are "caught" by their guru and carried across the landscape. They cannot predict when or even whether the guru will arrive; once he has gone, they do not know if he will ever return. They can control only the conditions—the quality of their attention and the purity of their intention—needed to invite the guru's presence. When he does appear, all the healers can count on for help in handling the danger inherent in the journey is their own skill and insight—their own presence of mind, body, and spirit—in the moment of the performance. There is no structure, no institution, nothing outside themselves to ensure that they will find their way back or that they will not lose their mind after their return. Any security attained from the ritual is similarly short-lived. It is only in the moment that things seem tame.

Village healers risk opening themselves to what is truly wild not because they are reckless or because they no longer care about the material world. They do so because, through their journeys into the realm of the ancestors and their ability to "see double," they have come to realize that things that seem solid and secure are nothing of the sort. Like the ancestors, they understand that the amount of grain in the grain bin or the size of the bird thigh has little to do with the essence that keeps us alive. The wisdom of the priests and the shamans—a wisdom that is healing—is their ability to recognize what they can control and to give themselves, utterly, to what they cannot.

13

MAPPING POWER

October 1992

A gray-haired Brahman sat at the edge of the darkened room, as far from the fire as he could and still feel its warmth. He lived in a village up the ridge; he was on his way south and had stopped for the night to speak with Devimaya's father, Dilli Prasad. Dhanmaya served plates of steaming rice and greens to her children, who were gathered around the fire. Dilli sat cross-legged on a straw mat beside the shrine for the ancestors, sipping *jad* and listening to the Brahman, who was talking about a meeting that Khagendra, a community-development officer for the Makalu-Barun Conservation Project, had organized the previous week to introduce the new park and conservation area. Khagendra had shown the villagers a map drawn on a large piece of white cardboard delineating the national park, colored green with Magic Markers; the conservation area, left white; and the boundary between the two, drawn in black. He then explained potential land-use changes in each area. In the green area, the Brahman thought, people would no longer be able to graze their cows, yak, and sheep. He could not remember what had been said about the white area, and he wanted to know what Dilli had heard.

Both men had pastures on the ridge above Hedangna on the way toward Makalu. Neither had seen a map like this before, and they were struggling to understand the relationship between the white and green areas, drawn on the piece of cardboard, and the trails and ridges of their home. As far as they could tell, if their pastures fell on one side of the black line, they would retain ownership rights to the land; if they fell on the other, they would lose those rights, a loss for which they were unlikely to be reimbursed. But neither had any idea where on the land the black line actually fell.

On the pilgrimage to the Khembalung caves near Makalu several months earlier, I had traveled through the same high pastures. After climbing through the rhododendron forests above Hedangna, the trail opened into small grazing areas, with bamboo shacks built on the perimeter, divided from the next grazing area by a stretch of forest. As we walked, a fifty-year-old man told me who had owned and used these pastures over the generations. As with the lower fields and forests in Hedangna, rights to these lands depended more on the relationships among users, the time of year, and the resource in demand—grass, dead wood, medicinal herbs—than on any rigid rules of land ownership. Like bargaining over prices in the bazaar, these rights were negotiated between owners again and again. The borders of a claim were always more vulnerable to encroachment than the center, and rights became more flexible with increasing distance from the village. These holdings might be represented on maps as a series of overlapping clouds, fading out at the edges, rather than as the discrete blocks bound by thin black lines on Khagendra's map.

For villagers who depended on livestock for their livelihood, the significant resource in these pastures was grass; the right to graze their water buffalo and cows was what they bought, leased, and sold. In creating a national park and conservation area on these lands, the international conservation community and the Nepalese government had declared that other resources and uses were valuable: biodiversity, tourism, cultural preservation, and sustainable development. These resources and uses required different sorts of ownership rights and different sorts of maps than did traditional claims to the land.

One day after returning to the village, I walked on the main trail toward a small stream to wash. I was well into my research by then and had notebooks filled with information about headmen and taxes and land disputes. I had elaborate kinship charts and carefully translated ritual chants. I had spent the morning alone in my room, going through my notes, and I was still so preoccupied with my struggle to understand the relationships between the different types of information that I was scarcely aware of the path beneath my feet.

I heard something down the hill. As I looked up to see what it was, I noticed a small section of fencing along the edge of a field that bordered the trail. The whole field had not been fenced in, only the area where wandering livestock might be able to reach the ripening rice. The farmer had decided what to do on that plot, based on his experience farming that piece of land. He might do something else in another field, and he might change what he had done here if it did not work, a decision based on what he observed on the land, not what he knew in his head.

This way of working the land—like the villagers' ways of marking boundaries on the land—did not lend itself to any abstract theories about farming or land ownership or life in rural Nepal. I was the one trying to make connections among categories I had created. My questions about how villagers related to the land and the natural world suddenly took on an abstract arbitrariness that reminded me of the discussion that Dilli and the Brahman had had over the green and white areas on the cardboard map. Khagendra's map was a set of lines drawn in bird's-eye view. This map was then laid across the landscape, with no way of accounting for the ridges and valleys that obstructed the view, for the rivers that made walking impossible, for the contours and details that gave the land its meaning.

By naming objects, we organize and make sense of a world that would otherwise be unknown, chaotic. Drawing maps and demarcating boundaries, our traces on the land, are ways of illustrating the particulars of naming and knowing, ways of demonstrating spatially what we see conceptually.

Naming the landmarks in the places we live is a way of entering into relationships with those sites. When we are forced to live where the localities and boundaries—and so meanings and values—were named and drawn by others, our relationship with those places shifts. This dislocation was most dramatic and devastating in regions colonized and governed by outsiders. A more subtle estrangement, a kind of "conceptual trapping," occurs when people are forced to live in a landscape according to maps and boundaries defined by someone else ("Encompassing Web" 1992:154).

◊

Khagendra's presentation in Hedangna in October 1992, which was repeated throughout the conservation area by other project staff, was the first step in introducing the Makalu-Barun Conservation Project to the inhabitants of the upper Arun Valley. He emphasized that the project would bring tourists to the region, which, in turn, would bring income-generating opportunities. He described how the project would do everything it could to preserve local culture, offering small grants to train villagers in *thankha* painting and drum making, to improve monasteries and village shrines, and to upgrade the schools. He said nothing about conservation or biodiversity because he knew that those concepts meant little to the villagers.

To Western and Nepali scientists and social scientists, the Makalu-Barun Conservation Project was about protecting the environment, promoting the culture, and developing the economy sustainably. Although these goals would be difficult to achieve, at least the tools for assessing the causes and extent of environmental damage and for proposing solutions were fairly straightforward. Anything to do with cultural traditions, however, was tremendously complicated. For example, one of the stated objectives of the park was to create a greater awareness among villagers of the importance of their cultural heritage. While the content of the programs to promote local knowledge and culture may well achieve that goal, their design does not. Educated outsiders conducted the research on which these programs were based, and then they—rather than traditional healers, farmers, and Buddhist lamas—introduced the projects into the communities. Thus the preservation of local traditions depended on the expertise of outsiders, not the efforts of those whose culture it was.

Similarly, like many conservation projects that pursue the perhaps fundamentally contradictory objectives of conservation and local development, at the heart of the Makalu-Barun Conservation Project was a gap between the rhetoric of local empowerment and the reality of project administration. The project called for increased local participation, yet regulations gave the government authority over decisions as minor as whether one tree could be cut down. Villagers had to make their own mistakes and learn their own lessons. If not, how would they ever feel empowered?

In Hedangna, the contradictions between the vision of empowerment and cultural preservation and the details of implementation were even more complex. One of the core values expressed in the Yamphu culture and sustained in *kipat* was that they were kings on their own lands. With the designation of the Makalu-Barun region as a park and conservation area—one of the first large-scale international development projects in the region—the national government in Kathmandu acquired more control over the lands of the upper Arun Valley than it had ever exercised.

This was a huge issue. The whole political history of the village had been oriented around maintaining control over the land. All the stories about the land that I was told started with the fact that the king could not "touch" *kipat* land. And yet, this internationally funded project, which intended to protect and promote local culture, knowingly or not was directly undermining one of the core values of one of those cultures.

Often the call to preserve cultural traditions as a facet of conservation projects means promoting the cultural traditions that further the objectives of outsiders and ignoring or undermining those that do not. For the Makalu-Barun staff, the project was about the worthwhile goals of protecting culture and the environment. But for the Yamphu—given the history of their interaction with the government and outsiders, given the reality of implementation—the project was ultimately about power, about the flow of money and resources needed to accomplish the broader goals, goals that the villagers did not really understand, even when they were explained to them. Any smaller programs to promote local cultural traditions and support the school were interpreted within this larger context of making decisions and exercising power.

During Dilli and the Brahman's discussion of whether they would be compensated for losing pastures, Dilli had suddenly shaken his head, as though just realizing something. "That is what the school grants are for," he said. "To make up for taking away our grazing rights."

◊

Kipat defined relationships between people, not between people and the land. The villagers' ability to enter the network of affiliations that gave them access to land, their identity as *kipatiya*, depended on having a *tsawa*. The *tsawa* identified a connection with the spiritual essence of the land, a connection that could not be seen but that made possible one that could be. In other words, access to the most important material resource in the villagers' lives—the land—depended on maintaining relations with the invisible domain of the ancestors.

The intersection of *kipat* and *tsawa* emerged from a worldview that did not distinguish between the political and the sacred; maintaining a relationship with each was necessary to secure and maintain rights to the land. The *kipat* and *tsawa* structured these relationships, *kipat* with the national government and the *tsawa* with the place itself, with the land and the ancestors of Hedangna. Each of these concepts spelled out what villagers must do to keep that part of their relationships intact: offering chickens and *jad* to the ancestors, offering taxes and *jad* to the headmen. None of these actions secured these rights for good. The villagers knew that they had to make these offerings again and again. It took work to maintain the relationships that secured their rights to the land they depended on to live.

At the heart of their connection to the land was their relationship to the ancestors and the sacred, to a landscape infused with invisible forces beyond their control. Instead of trying to block these forces from their lives, villagers named them—Chaketangma and Matlung Thuba—and fed them chicken and millet. They invited them into their homes, building a shrine for Khammang (Grandfather) and Yimmahang (Grandmother) on the eastern wall of the house, adjacent to the cooking fire—a constant and visible sign of "the world they did not make" (Cronon 1995:87)—which reminded them of the limits, not the extent, of their control. Villagers remembered the ancestors because they knew from experience that if they ignored or forgot them, children would fall ill and crops would fail. They knew, conversely, that if they honored the memories of their predecessors, families would be healthy and grain bins full. There was nothing abstract about it.

This attitude—the relationship to the land expressed in the *tsawa* and *kipat*—was what was most threatened by the management plans of conservation. The day after Dilli and the Brahman's conversation, Ganesh, Amrit, Chute Rai, and other men were sitting on Ganesh's porch, drinking *jad* and discussing the park. They were also talking about whether grazing would be restricted and, if so, whose pastures would be affected. I asked what they thought about the park. They all shook their heads. "This land is our home. *Hernos, bahini,* look, little sister," Chute said, waving his hand across the land. "The rocks and trees are like objects inside our house. This project is like having someone come into your home and tell you what you can and cannot do in a place that has always been your own."

Like similar efforts in other parts of the world, the Makalu-Barun Conservation Project set out to support local culture along with promoting biological diversity. The reason for the existence of these projects is premised on reestablishing links between nature and culture, between humans and the earth—connections that have not yet been severed in Hedangna. The restrictions imposed on resource use have been based on models developed by scientists and social scientists, not on a relationship with the ancestors that, in turn, depends on acting with restraint and respect. The project aims to control the environment, not enter into a relationship with an entity that can never be controlled.

After reading about the politics of conservation in developing countries, I was not surprised at the displacement of power created when a well-intentioned and well-funded outside organization became involved in the management of local resources. Yet I also began to sense that in this shift, something deeper was lost. Since the land in Hedangna has been surveyed and *kipat* is no longer in operation, perhaps this deeper shift is ultimately the more significant change in the villagers' relationship to the land. Ideas from the outside—even good ideas—are different from those that emerge from within a particular landscape and culture. It is not only that there is a depth of understanding in the villagers' relationship with their land that outsiders often fail to grasp. When we begin to trust or are forced to trust someone else's decisions about what is best for our land more than we trust ourselves, we risk losing something important, an internal integrity and resilience, the ability and trust needed to take care of our land using our own ingenuity, powers of observation and affection, and capacity for restraint. We risk becoming dependent on the vision and expertise of others to see and do what we begin to sense we cannot see and do on our own.

14

LOST SOULS

January 1993

I walked quickly up the ridge so I would arrive while it was still light enough to see the trail. When I reached the house, Amrit, who was sitting outside on the porch, immediately asked if I had come over the top of the ridge and repeated what I had been told before: never walk alone near the ridge at dawn or at dusk because that was when Matlung Thuba wandered the hills looking for souls to snatch. If Matlung Thuba got hold of my *lawa*, he said, I was likely to die.

This was what people suspected had happened to Amrit's brother-in-law, Chandra. A month earlier, Chandra had gone to his fields just before dawn. He had a fever when he returned home several hours later. He stopped eating and after a few days was unable to talk. Soon he could not recognize anyone, not even his wife or four sons. His family called shamans and priests from everywhere around. Chandra simply got worse. Tonight, they had called the

Gadi shaman, considered by many to be the most powerful healer in Hedangna. Amrit said that they were also making plans to carry the sick man to the hospital in Khandbari.

◊

Amrit was Ganesh's neighbor and cousin. During my first weeks in Ganesh and Jaisita's home, Amrit had come to the house almost every evening after eating. He would chat with Ganesh and Jaisita a bit before turning to me to ask a series of questions, always on the same topics: New York City and Michael Jackson. In the morning, before or after eating, he would come to my room to look at my belongings. He was one of the first villagers to do so and the one who returned most often to look at them again. He asked about Westerners and whether I thought he could get a job as a cook for a trekking expedition. Sometimes he brought Ganesh with him and asked me to teach them English phrases such as "Would you like tea?" and "Can I be your porter?" that had a use, not like the sentences in Devimaya and Deuman's textbooks, which made grammatical points.

Amrit asked many questions, but he never seemed very interested in my answers. After asking one, he moved to the next or became distracted by my camera or my stove. Once, after staring for several minutes at the books lined up against the wall of my room, he asked whether I had had to pay money for each one. He was as surprised by my affirmation as I was by his question.

Amrit had a broad face that made him seem chubby, but no man in Hedangna was actually chubby. He wore a blue T-shirt and pale blue nylon shorts. His father was Ganesh's father's brother, but Ganesh's father was "clever" and had managed to increase his landholdings far beyond what he had inherited. His father was not clever, Amrit said. Plus, Ganesh was an only son, and Amrit the youngest of three, which made the difference in wealth between them even more extreme in a community where property was divided equally among sons. To me, both Ganesh and Amrit seemed like young men who wanted to stay out late and have a good time, neither ready to assume the responsibilities they had acquired as husbands and fathers. Ganesh, at least, had his wealth to fall back on. Amrit had only Ganesh.

Amrit was the most playful and spirited man I knew in the village. During the ritual calling for rain and a good harvest the previous year, I was sitting with a group of men on the hard-packed fallow rice fields, waiting while the village priest finished the ritual for a neighboring group. The sky was clear, and the sun was beginning to slip behind the ridge. The late-afternoon air was cool, and I pulled my shawl around my shoulders. Amrit told me to dance. I protested. Okay, he would dance, he said, with no prompting, as though this had been his plan all along. He asked me to sing a disco song and said that he would show us how to dance disco. I did not know any disco songs, so I sang "Rocky Mountain High," the first tune that came to mind. I sang it fast so it might sound like disco, or whatever Amrit imagined disco to be. Amrit said that it was perfect, and he leaped into the center of the circle, where he began to dance around and around, performing his version of a disco dance. The men laughed and clapped and called for more.

◊

Despite his friendliness, Amrit triggered something in me that Devimaya and her family never did. As long as I knew them, no one in Devimaya's family ever asked for any of my possessions. This made me want to give them everything. I gave Devimaya T-shirts, lungis, and the one sweater I ever knit for myself. When I left, I gave her mother my Synchilla jacket and her father money.

People who envied my things triggered exactly the opposite reaction; their envy brought out my selfishness. Like his children and his mother, who lived in the house below his, Amrit picked up my belongings when he entered my room. This immediately put me on guard, and, even though I wished otherwise, I watched him carefully to make sure that nothing disappeared. Despite my apparent stinginess, he was warm and outgoing with me, until I refused to lend him money. I refused because I knew that he would never pay me back and feared the dynamic that the default would create—between Amrit and me as well as with others who would want to borrow money. I decided that it was best not to lend any money at all. After telling Amrit the reasons for my refusal, he left without talking. For the next few weeks, he was cool and distant and did not visit Ganesh's house as often as he had. Eventually, both of us seemed to decide to forget what had happened. Even so, this exchange marked the edges of our worlds, the difficulty each of us had in seeing the other as an other, not just as a symbol of the category we were taken to represent.

◊

The evening I arrived at Chandra's house, Amrit was working outside on the tiny porch with several other men. Sidhiman, Raj Kumar's father and Chandra's father-in-law, placed branches against the mud wall. Another man shaved bamboo stakes with a *kukuri* for the shaman's wands, while others filled plates with piles of rice. Amrit told me to go inside, that they had work to do, and I would be in the way.

I ducked under the low door and entered the dark, smoky room. Chandra was sleeping on a mat by the fire. The shaman sat on a long straw mat by the dying flames, drinking a *tongba* and talking to the women. He wore a gray, button-down, store-bought shirt; what looked like green army shorts; and a vest. Chandra's mother sat close to the shaman, tending the fire, pouring more water into the shaman's *tongba,* and listening to what the shaman had to say. Chandra's wife, Raj Kumar's oldest sister, sat cross-legged on the mud floor, leaning against the wall, her youngest son nursing at her breast. Her hair hung loose and tangled to her waist. Her face and clothes seemed to be covered with a layer of dark grime. She stared vacantly into the distance.

After a long time, the shaman was called outside and began to prepare for his journey. He put a white cotton Nepali tunic over his shirt, took off his shorts, and put on light-cotton Nepali pants. He hooked a belt of bells around his waist and hung a betel-nut necklace around his neck. He wrapped a handful of leaves in a towel and then tied the towel around his head as a headband. He did each action slowly, precisely, with a lot of discussion along the way. After a long time, he picked up his bamboo wand and signaled to the drummer that he was ready to begin.

Sidhiman, sitting by the shaman's side, began to bang on a bronze plate. The shaman slowly and not very energetically began to chant. I sat close by, inhaling the sweet smoke of burning juniper and the musty odor of urine and absorbing the repetitive clanging of the plate and the deep tap, tap, tap of the drum, which had been heated by the fire so the sound would be deep and strong. The shaman kept stopping and then starting again, only to stop. Just when I thought that he finally had begun, he stopped. He told the men to bring the sick man outside. Chandra's younger brother went inside, carried him out on his back, and lay him on a straw mat beside the shaman. Sidhiman, whose manner was usually abrupt and gruff, placed Chandra's head in his lap and wrapped his arms around his son-in-law's shoulders. Sidhiman's daughter, Chandra's wife, came outside and sat on the doorstep, her four sons by her side. Again I was struck by the empty look in her eyes.

Even in the presence of the sick man, the shaman did not do much. He sat in front of the altar, moaning out his *mindhum*. At one point, he did stand up and dance around a bit. But then he stopped and began to talk. Amrit yelled at him to stop talking so much, to stop wasting their time. "All these people have come to see you!" he shouted. "We need to get to sleep! Now do your work!"

The shaman again began to dance, but then he fell and had to be helped up. At another point, he tripped and almost fell. Finally, the shaman announced that a young goat had to be sacrificed. Usually, everyone immediately set out to prepare a sacrifice called for by a shaman. But by this time, the tone of the performance had changed. At the beginning of the ritual, many neighbors had arrived to support the family and to help out as they could, but most had gradually drifted home, leaving only the immediate family and close neighbors. There was some talk about goats; one was finally found and was sacrificed. But instead of cleaning the goat and cooking the meat to feed the onlookers, Amrit and the others just left it lying headless in a puddle in the rain. The ceremony was over. The shaman went inside the house to sleep.

Sometime later, Amrit joined several of us who had gathered around a fire under a small shelter next to the house. As he walked toward us, he picked the goat out of the puddle and threw it onto the woodpile with a look of disgust. He sat down, and I asked what had happened. "The shaman hadn't met his guru," he said. "That means either that Chandra is going to die or that the shaman has had too much to drink. Usually," Amrit explained, "this shaman danced all over the place. In one ritual, he even danced all the way up from his own house in the Gadi to the home where the performance was to be held. But this time he didn't do anything. He was too drunk."

Several days later, Chandra's parents took him to their home across the ridge in Uling, where they had another shaman perform a ritual. The next I heard, they had taken him to Khandbari, a hard three- to four-day walk when carrying a sick person on a stretcher. He died on the trail, several hours above the hospital.

◊

Shortly after Chandra's death, I stopped by Chute Rai's house one afternoon after planting potatoes with Raj Kumar's wife in the fields below his house. Chute was huddled under a woolen blanket by the fire when I entered the dark room. He weakly pointed to a straw mat for me to sit on. He sat up slowly, pulled the blanket over his shoulders, and moved closer to the fire. He stared into the flames, stoking the coals. I had never seen him so despondent. The fever had returned during the night, he said. A deep, hacking cough interrupted his words and shook his body. He finished, finally, and stared silently into the coals.

He had no strength anymore, he continued. He did not think that he would live much longer. He sighed. He looked up at me and then back at the fire. All that the priests and shamans did was make the villagers spend money, he said. But they had no other choice, and so he had killed chickens and goats and pigs, and still he was no better. He listed all the people who had died in the past year. "People should live longer then this, *hernos*," he said. "We should live until we are seventy, eighty. But we don't."

Although the ancestors are responsible for creating life as the Yamphu know it, although they are the ones that bring good fortune and rain, the most common encounters the villagers have with them are through illness and death. Every person has four or five different *lawa*, the Yamphu concept for soul or essence of life, any of which might suddenly be snatched away by an ancestor or lured away by the soul of someone whom the person loved and who has died. The soul of a strong and healthy person is unlikely to leave her body. But if a person is startled, sad, or just worn out, his *lawa* could become lost, either following those who have died into the realm of the dead or being taken by an ancestor, who would then demand a chicken or goat to ensure the soul's return.

A person with a fragmented or lost soul becomes listless. She falls ill. A shaman is called to diagnose the cause of the illness and heal the sick person by ensuring that her soul is again intact. If a healer fails to connect with his guru, the source of his power, or if he fails to reach the pool at the center of the world and experience the flash of insight into the cause and cure of the illness, he will be unable to recover the lost soul. For the Yamphu, when a soul cannot be retrieved, the body will die.

◊

Amrit's mother, Kalimaya, died a few weeks after Chandra's death. The morning of her funeral, held three days after her death, was overcast. Heavy, gray clouds lay on the ridge across the river, and a light dusting of snow from the previous night's storm still covered the pastures on the ridge above Hedangna. It had poured all night in the village, and the air was moist. I walked to the edge of the terraced rice field, where twenty or thirty men huddled around several small fires to keep out the raw January weather. They wore woolen scarves wrapped around their heads and torn cotton blazers or ragged, factory-made sweaters. I squatted between Chute and another man, who made room for me, and reached out my hands to draw some warmth from the flames. I listened distractedly to comments and opinions I had heard before and looked around to see who was there and what they were doing. Next to the fires, younger men spread steaming rice from huge brass containers on flat bamboo mats to cool. They chopped cooked buffalo into bite-size pieces and boiled the blood into gravy.

Sometime later, a slowly moving line of color filed across the dust-colored land; women, woolen shawls over their heads, were carrying wooden containers filled with fermented rice or millet. The faded reds and greens and blues of their clothing were luminous in the flat light. They went beyond the men and the fires to the far side of the field, near a stream. Sitting on the ground in small groups, they mixed the millet with water in large tin jugs to make beer and wove twigs through large leaves: plates for the meal of rice and buffalo that would be served after the ritual part of the funeral was complete.

Ganesh and Dilli laid two handwoven, brown-and-white woolen blankets over the spiky rice stalks still standing from the fall harvest. Five *tongbas* were placed on the ground next to a pile of rice and some coins. Five older men were called over, and each sat cross-legged behind a *tongba*, which was pay for their help in performing the ritual. When everything was set up, Amrit and his brother Ratna came over and sat down on the blankets.

I had stopped by Amrit's house the day after his mother died to say how sorry I was and to bring him some packaged soup I had brought from Kathmandu. Amrit was sitting next to his brother; he quickly averted his eyes when I came to the door. A neighbor shook her head and told me to leave, that I was another caste, and I should not talk to Amrit until after the funeral.

Amrit and his brother were naked under the large, white homespun shawls they had been wrapped in when I had visited his house, and they had wrapped white cloths around their recently shaven heads. Seeing Amrit now in nothing but a woven shawl, I was struck by how thin he really was. His eyes were hollow, with dark circles underneath.

The five men began to talk in loud voices, although only the adjacent crowd paid any attention. One man spoke, tentatively, as if asking a question or offering a suggestion. Another interrupted, and then the first responded. They were relaxed; the scene was informal. They spoke in a mixture of Yamphu and Nepali. Chute Rai started, "Today, we, the five clans [*panthar*], will purify the land from the death of Kalimaya." He paused to sip his *tongba*. "Lo," another man added. "Impurities are attached to the sun and impurities are attached to the moon." And then another spoke up, "Consecrated rice, make them pure." Chute Rai again took over: "There are impurities in the hills and impurities on the ridges; there are impurities in the earth and in the soil. Consecrated rice, make this pure."

They called on Kalimaya's soul to take the impurities with her. They told her to use the butter lamp to light the trail north to Lhasa and beyond, where she would be carried up a ladder into heaven, and to use the money if she was stopped. They instructed her to drink the *jad* when she was thirsty and to eat the rice when she was hungry. Amrit, who had been quiet until now, spoke up, "From now on sing songs. Dance!" And then Shyamlal, another of the five men: "We, the five clansmen, have made you pure. *Now go!*" Together, Amrit and Chute concluded:

> *Don't give us trouble.*
> *From today, it is your holiday;*
> *Don't come back.*
> *From today we take leave—it is over.*
> Khardey! Khardey! *Now go. Go!*

◊

Kalimaya had been in her seventies. When I had first arrived in Hedangna over a year earlier, Kalimaya, like every other woman in the village, had headed off each morning with her bamboo basket to the fields and forests for a full day's work. Then, three months ago, she began to lose energy. At first, she complained that her skin, her back, her legs, and her arms itched all the time. She lost her appetite; she no longer felt like eating rice or even drinking millet beer. Her face became gaunt, and her skin took on a yellowish tinge. She huddled outside in the sun or crouched inside by the fire, trying to escape a constant chill. Her sons called priests and shamans to learn what was wrong. They killed chickens and goats and even a small pig. Nothing worked. People shook their heads; they whispered that she had been poisoned and that she was certain to die.

Amrit and Ratna finally decided to take their mother to the nearest hospital in Khandbari, a hard twelve-hour walk from Hedangna even without a load. Amrit hired a low-caste laborer for 500 rupees, enough money to live on for several months, to carry Kalimaya in a basket on his back. The porter had demanded 650 rupees, but by promising to take turns carrying his mother, Amrit had convinced him to accept the lower pay.

After three days, they reached the single-story hospital on the southern edge of the Khandbari bazaar. When I had visited the hospital, women bringing children for inoculations and medicine sat in the hard-packed mud courtyard, nursing their infants while waiting their turn in a line that stretched across the courtyard. Occasionally, they lifted a hand to swat away flies. The stone walls were crumbling, and the whitewash was chipped and cracked, exposing sections of rock underneath. Inside, the hospital was dank and dingy. Voices echoed off the walls, making the space feel even more crowded than it was. It smelled of sickness and sweat.

Kalimaya was given liquid glucose and one of the few empty beds. A doctor said that she had hepatitis and that the only cure was rest and good food. Her appetite returned while she was in the hospital, Amrit later told me, but her bed was needed for more serious cases. Amrit carried her for three hours to his sister's house, in a village on the ridge just north of Khandbari. She lost her appetite again, but she refused to be carried back to Hedangna until she was stronger. And then one morning, she stopped breathing.

Most of the villagers who had come to Kalimaya's funeral sat around the fires and talked or continued to cook and prepare the food. Only a small group of closer relatives and neighbors sat clustered around the five men who had been called to purify the village and to ensure that Kalimaya's soul did not linger in what was no longer her home.

Tulimaya, Ganesh's mother, who was ten years younger than Kalimaya, sat on the edge of the woolen blanket on which the men performing the ritual were seated. Tulimaya and Kalimaya had married brothers and, in a village where brothers lived next to each other, had become neighbors as well as sisters-in-law. They had raised their children side by side, collected wood in the surrounding forests together day after day, met at the water hole each morning, and shared millet beer in their dark, smoky houses. The two oldest women on the pilgrimage to the sacred caves at Kembalung, they had hiked for four days up to 15,000 feet at the base of Makalu, the fifth highest mountain in the world, and then climbed for an hour up a rock face to make offerings to Shiva. Each had burned her husband's corpse on the banks of the river below the village.

Now Tulimaya watched the men purify the family, the clan, the soil, the hills. She wore a dark green satin shirt, untied at the neck and torn at the wrist. Her black hair, streaked with silver, hung long and tangled down her back, and her maroon shawl had slipped down around her shoulders. She wiped her veined, bony hand across the tears slipping down her cheeks. She stared out across the fields; her eyes looked lost and alone.

Kalimaya had been my neighbor for ten months. She had stolen my matches and had borrowed my gray pile jacket for the pilgrimage to Kembalung and returned it unwashed. Eventually, she tried to teach me the Yamphu language. She showed me how to clear out saplings with a sickle to open land for planting millet, and she called me into her home for bowls of warm *jad*. The previous summer, I had scrubbed her bony back with a rock to clean off the dirt from a month of planting rice. After I moved in with Dilli and Dhanmaya, I saw her less frequently; when I did pass her house, she would call me over to sit next to her in the sun, to talk to her a bit and help pass the time.

Later in the day of Kalimaya's funeral, I carried a metal jug in a bamboo basket to get water from the stream. I took the long way, walking slowly through the brown stubble left over after the rice had been cut. It was not only Kalimaya's death that made me melancholy and lonesome. The ridge across the Arun River faded into the thick, dark clouds that had hung heavily over the village for days. The sky was white; the land, stark and barren. The food had been the same oversalted spinach and heap of rice for as long as I could remember. The air made it too cold to wash. In a city, I could distract myself from this emptiness. I could move on to other things and, in the movement, begin to forget.

Tulimaya could not escape it. Each day, she would walk by the house where her neighbor had cooked rice, argued with her grandson, and picked the lice out of her shirt. After eating each morning, she would have to set off alone to the fields or forest to do the tasks that she had done with Kalimaya. The pain would grow less acute, but not by walking away from the places where that pain was lurking. And, now, in winter, the land reflected the starkness of death, the chill that was impossible to escape. Warmth would come only with the passing of time.

The night after the funeral, Dhanmaya told me that a village priest had told Amrit and Ratna not to take Kalimaya to Khandbari. He had said that they should leave her in Hedangna to let her die in her own home. But, Dhanmaya said, Amrit is clever and skilled. He did not take others' advice; he had carried his mother south to Khandbari, and she had died like someone with no sons. She was buried in another land, away from where her mother, her father, her husband, her brothers and sisters, and her children had died.

A neighbor had stayed at Amrit's home the first night after he returned to Hedangna without his mother. He later said that Amrit had sobbed, loud shaking sobs, all night long. But it was not only her absence that made him sob. In the days following his mother's death, Amrit said that he could not sleep at night. Because he had ignored the advice of the village priest and had tried to prevent one pain—the pain of separation—he had created another, a deeper and more enduring pain. According to Yamphu oral tradition, villagers should be buried or cremated in Hedangna, with their own people on the lands of their ancestors. It was too far to carry Kalimaya's body home, so she had to be buried in another country, another *des*. For Amrit, the land of Hedangna was now even more potent for what it lacked—the absence of his mother's body—and for the blame embodied in that absence.

A place can hold roots; it can provide a sense of belonging and a kind of grounding. A place can also become a trap filled with voices and memories that haunt those who try to live their lives there. In Hedangna, the villagers cannot keep out the past; they cannot even keep out the cold. They literally live the successes and failures of their ancestors, whose fields they till, plant, and harvest year after year. For them, home is not a location that is chosen, and a "sense of place" is not an emotion they decide to nurture. Home means staying put—staying put with life, with death, with the cold and the heat, with living next to neighbors who may be your relatives and also your enemies. Home means building the present on the landscape of the past.

PART THREE
return

15
LEAVING

February 1993

A few weeks after his mother's death, just before I left Hedangna after eighteen months of research, Amrit announced that he was moving his family to Khandbari to start a tea shop. Amrit did not have enough land to support his family, and the illness and death of his mother forced him even further into debt with Ganesh, his cousin. He was going to try something new, he said. Amrit had an almost insatiable curiosity about the world beyond the village, and I was not surprised when he decided to move. But curiosity was not enough to survive in this world, and I wondered what would happen.

The week before they moved, the family called the priest Purnamba to their house to ensure that their souls were healthy and intact before leaving the village. Amrit invited me to come. When I arrived with Kumari after dinner, the room was crowded with neighbors and relatives who had come to offer support and advice before Amrit and his family left. We squeezed into an empty spot near the door and joined the others sitting cross-legged on the mud-packed floor. I looked around the room to see who was there. I noticed Chute Rai, Ganesh, and Kharka Bahadur, with his two crooked, decaying front teeth. Long shadows accentuated their angular cheeks. These men looked tired. They had seen these ceremonies before. They came because they were brothers and neighbors and because Amrit's mother had died and he was leaving the village. Amrit would have done the same for them. They turned to me and asked if we did this in my country.

Purnamaba, wearing a short white tunic, was standing before an altar. His dark, stocky legs were bare. He clutched a clump of leaves in one hand and two bamboo wands in the other. He closed his eyes and began his chant. He shook the wands and branches, his chant became faster, and his body began to shake. He stopped abruptly and turned to Amrit, who was sitting cross-legged on a woven mat surrounded by children. Purnamba raised and lowered the wands over Amrit's head, chanting loudly and then softly. His eyes were closed. Everyone else spoke noisily, not paying much attention. Only Amrit was silent, his head down, hands resting on his folded legs. After several minutes, Purnamba stopped short and began to speak. The room became quiet as the participants paused to hear the diagnosis. Kalimaya had taken Amrit's soul, Purnamba said briefly and matter-of-factly, as though this were what he had expected. He turned to the altar.

Purnamba again chanted and danced as he searched for signs of the lost soul. After a few moments, he stopped. He turned, placed the leaves on the altar, and went to Amrit's side, keeping his empty hand tightly shut. He chanted some more, waved the bamboo wand with his other hand, and then blew into each of Amrit's ears and onto the top of his head. The shaman explained what he had seen. Amrit's soul successfully recovered, Purnamba stepped before the person at Amrit's side. Those who had paused to hear his account of his travels returned to their gossip. The children drifted back to sleep.

After checking on the soul of everyone in the room, Purnamba sat down cross-legged by the dying fire. He started to talk about Amrit's plans for the family's move. Amrit mentioned that a Brahman from a neighboring village had consulted the astrological chart and advised them to leave the following Wednesday, four days earlier than Amrit had planned. Purnamba told them to follow the Brahman's advice. He told them to perform rituals to Yiwa, the god of the Arun River, before leaving the village and at the shrine for the Devi above the *tsawa*. And he told them not to cry when they left.

I listened as this man, who had never been as far as Kathmandu and had never lived anywhere other than in this village, offered counsel for the family's move. He talked about what he knew, advising Amrit and his wife about how to leave Hedangna correctly and conscientiously. He had little to say about what to do once they arrived in Khandbari, which was beyond the realm of his authority—the Yamphu ancestors did not travel that far. Down there, Amrit would be on his own.

As they talked, Amrit and his family wove bits of string into small necklaces. Before the altar, Purnamba mumbled another low chant, retracing the original journey of the ancestors to Hedangna, and asked the ancestors to protect those who wore the necklaces. He tied the blessed strings around the necks of Amrit, his wife and sister, and the three children, and told them to wear them for six months.

It was late. Kumari nudged me, and the two of us stood up to leave. Amrit followed us into the moonlight to say goodbye. He leaned against the stone wall and said that his sister kept telling him to move quickly, that he had to get to Khandbari to buy firewood, fermented millet, rice, and meat. His sister, a widow who lived outside Khandbari, the only Yamphu in a Lhorung Rai village, wanted him to move, he said. His mother was buried near where his sister lived, and his children could get a better education in Khandbari. "But, who knows. If it weren't for all this, I wouldn't leave," he said pensively. "I would stay here in Hedangna."

As I listened to Amrit in the moonlight, I thought about what he was leaving behind. I thought about the loneliness that I had felt while living in the village, especially during the first few months, and wondered if Amrit had ever experienced such feelings. He had no relatives in Khandbari. Only one other Yamphu family lived there, and they were from a village over the hill, not from Hedangna. I wondered how it felt to leave a community that was more tightly interwoven than anything I had ever known. I wondered what he feared the most.

He paused and looked into the long, bright shadows cast by the full moon, gazing past his brother's thatched-roof mud house, only a few steps from his own; past Ganesh's home; and into the fields where the potatoes were beginning to sprout. He would try it for a year or two, he said. "If things in Khandbari don't work out, I won't come back. There won't be anything to come back to. I'll move on to something else."

Amrit said goodnight and turned back into the smoky firelight. He sat up most of that night and the next, talking with his neighbors, his family, and the priest. They spoke about selling the family's remaining stores of millet and corn. They discussed what it would be like in Khandbari and when they would meet again.

◊

A few days later, I was squatting on the banks of the Arun, this time for a ritual to mark my own departure from Hedangna. Baiseti Thuma sat on the bank, along with Dilli Prasad, Chute Rai, Raj Kumar, and many other men and women I had come to know in the village. Baiseti had walked for an hour and a half down the steep trail to the banks of the river, farther than she had walked in years, because I had invited her to come, she said, and because I was leaving and the villagers had to do what they could so that I would pass.

It was early in the morning and cold, and my hair was still damp from bathing in the icy water. The sun had not yet reached the valley floor, and I shivered. The water was blue-white from the glacial silt eroding from the mountains to the north. Men shouted over the rapids. Women crouched on the rocks above.

I clutched a scrawny chick that was for Yiwa. Amrit clutched a chick. Ganesh clutched a chick. Raj Kumar and Dilli held a goat. The shaman Kelekpa turned and told me that this was my ritual; it was in my name; it was for my good fortune. I had to think about what I wanted and to think about it again and again while he called on Yiwa. He turned back to face the river and began to chant. He waved a brass urn filled with weedy flowers. He continued to chant and then began to shake.

I looked at the river. The ritual, I thought, was a way to celebrate my time in Hedangna and to thank the villagers for everything they had done. I planned to ask Yiwa, the most important of the Yamphu's gods, to give strength to the community in light of the changes coming their way: the land survey, the national park, the hydroelectric dam, and the road that was to come with the dam. I stared at the water as it rolled over and around huge boulders, pulling down logs and leaves, transfixed by its roar and its force. I felt the tiny bones of the chick squirming in my hands. I took a deep breath as I realized that I was really doing the ritual for myself. I was concerned about the future of the villagers. I was also concerned about my own, although for different reasons. I asked that the openings I had experienced, within and without, did not slam shut when I returned home. I asked for the wisdom to write rivers, to write the words that carved out the valley and were carved by the valley; the strength to tell the story that was here to be told, even though the story would always fall silent before the thundering roar of the rapids.

◊

On my last afternoon in the village, Raj Kumar and I went to Chute Rai's house with a final tape of Purnamba's *mindhum* that still had to be translated. Chute was glad to see us and agreed, as always, to help. It was a cool, sunny day in mid-February. We sat cross-legged on the mud-packed porch, our backs against the wall, looking across the terraced fields, and painstakingly translated each word from the Yamphu ritual language into contemporary Yamphu and then into Nepali.

As we worked, the moon, full and bright, began to climb over the ridge above the Arun. It was huge. We stopped to watch it rise through the darkening sky. After watching silently for some time, Chute waved his hand toward my tape recorder and toward the notes that were harder and harder to see in the fading light, dismissing the work that remained.

"You have what you need, *bahini*, little sister," he said. "All the things that matter, all the things you have to know are here in the song we have translated: the *tsawa, kipat*, the *gaorung* [headman], the *mindhum*. You have it all," he repeated. "Here, in the song."

I knew that he was right—not that I had it all, but that I had what I needed. We turned back to watch the moon.

◊

I returned to Dhanmaya and Dilli's home late that evening, after Dilli and the children had eaten and gone to bed. Dhanmaya sat in the dark by the dying coals, waiting to eat until I returned. We did not say much as we ate rice and lentils. The silence was comforting, a silence of familiarity, not because we did not know what to say.

After Dhanmaya went to bed, I sat on a large boulder above my tiny one-room mud house. The sky was clear; the full moon, bright. I looked across the valley at the gray-blue ridge and the snow-covered peaks in the distance. I watched the bamboo move in the wind like a huge, dark animal and felt the silence of the thatched-roof houses, white mud walls shining in the moonlight, the steady flow of the river the only sound now that everyone was asleep. My clothes smelled of soil and sweat and wood smoke; the air smelled clear and clean, like the streams. I thought of the sameness of my life here, of the villagers' lives here—walking the same paths, planting the same fields, day after day, year after year. A sameness that could be a source of security or of drudgery, depending on whether I was filled by the experience of living in Hedangna or was empty and lonesome, missing the things I knew. It was a sameness that I knew I would long for, once I was gone.

I awoke early the next morning. My bags were packed. I had given away as much as I thought I could, and still my pack was heavy, weighed down by my clothing and notebooks, by my camera and tape recorder and sleeping bag, always too many things. I began making my rounds to say goodbye. At each house, I was fed fried eggs, potatoes, bowls and bowls of *jad*; some gave me sacks of pounded rice that had taken hours of work to prepare. All the food I wished I had been offered over the past eighteen months was generously before me to eat in a few rushed hours.

I sat on the stoop with Amrit's brother's wife and talked about the departure of Amrit's family. She said that she already missed them, and now I was leaving as well. "We see other foreigners, see them pass by on the trail when we go to Khandbari," she said. "But we can't talk to them. They aren't like you. You've lived here. You know what it is like; you know the *sukha* and the *dukha* [the good and the bad]."

I thought of Kumari's comment the night we had headed home from Amrit's house after the ritual. I had asked if she wanted to live somewhere else, such as Khandbari or India. She shook her head and said no, that she wanted to remain here, in her own country, her own home. "Do you ever want to live somewhere else, somewhere other than your own home?" she asked, forgetting, it seemed, that my presence in Hedangna meant that I was living somewhere other than my home.

I went to Chute Rai's house last before leaving Hedangna and lingered the longest. His wife brought more *jad*. We talked about what I would do in Khandbari, how long I would stay in Kathmandu, topics we had covered the previous day. I still did not stand up until finally Chute looked across the fire and said, "The day is getting on, *bahini*, younger sister. *Janne manche janna parcha* [People who are going have to go]." He repeated, "People who are going have to go." I knew that he was right. We stood up and went outside. I hoisted my pack onto my back. We clasped hands and held them tightly, each knowing as we held each other's eyes that this might be the last time we saw each other, knowing that if we did see each other again, it would be different. He stood on the edge of his courtyard watching me as I turned and walked down the path.

◊

I walked through the Gadi, down the steep hill to the Arun River, and back up the even steeper climb to Num, repeating Chute's words again and again, "People who are going have to go," a mantra to keep my feet moving across the rocky earth, the earth that was the villagers' home and not my own.

I spent the night in Num, just across the river from Hedangna, and set out early the next morning in the dark. After a ten-hour walk, I entered the outskirts of Khandbari, a completely different world from Hedangna. I walked past the stores and houses and schools and government buildings, finally entered the bazaar, and found Amrit's tea shop. His five-year-old daughter ran to tell him that I had arrived.

I had seen Amrit four days earlier by the bridge over the Arun, just below Hedangna. It had been the morning of his Arun ritual (*panchewlis*) for Yiwa to request good fortune for his family's journey and new life. He, his sister, his wife, and their children had bathed in the river and made offerings. After eating, they had loaded their baskets and headed off across the suspension bridge, up the long climb to Num, and on to Khandbari, their new home.

On the morning of the ritual, Amrit had been dressed in cotton shorts and the torn blue T-shirt he always wore. Now that he was a tea-shop proprietor in Khandbari, he wore his best clothes: a navy blue sweat suit with a yellow stripe down the side and a pastel cap. He invited me into his new home. It was dark and smelled of smoke. The ceilings were high, and the worn wooden floors creaked underfoot. Two windowless rooms opened off the dimly lit hall, which led to the back room. Devimaya had come from the nearby village where she was studying for her high-school-leaving exam to see me off. She was waiting in the back room and talking to Amrit's wife. I dropped my pack next to hers on the wooden bed in one of the tiny rooms off the hall and followed Amrit to the back of the house.

They were running low on fermented millet, Amrit said, but there was just enough for us each to have a *tongba*. I told him that I did not have to have a *tongba*, that they should save the millet. But Amrit insisted, saying that it was our last evening together and we had to celebrate. He filled several wooden containers with millet taken from one of the three or four bamboo baskets leaning against the side wall. The baskets held what was left of the rice and millet that the family had brought from Hedangna. They had already unpacked their blankets and clothing and had given away everything else.

We sat at the table and talked about Hedangna and what Amrit was going to do in Khandbari. We drank our *tongba* and ate rice and boiled pig meat. I bought a Star beer to give them a taste. Devimaya took a sip, grimaced, and went back to her *tongba*. Amrit, who had just moved to the bazaar where people drank beer, was less direct. "It grows on you," he said. We lingered, talking a bit more about what I would do when I got home and when I would return. But there was not a lot left to say, and Amrit soon said it was time for us all to go to bed.

I spread my sleeping bag on the bare wooden bed next to Devimaya's blankets, and we lay on the hard surface. A single fluorescent tube hung at a slant against the wall. We talked until drifting off to sleep. Sometime in the middle of the night, when everything outside was dark and quiet, I awoke to a loud crash, a cry of pain, and a low dull moaning. Devimaya jumped out of the bed and ran into the back room. I followed her and, in the darkness, bumped into Amrit's wife, who was fumbling to find the light switch. She finally turned on the bulb, perhaps the tenth time in her life that she had turned on an electric light, and the two of us followed Devimaya down the rough wooden stairs into the basement.

Amrit, dressed in only his blue nylon shorts, was squatting on the damp mud floor and clutching the back of his head. Blood seeped through his fingers. He moaned, crying out that he was going to die, he was going to die. Sahila, a villager from Hedangna who was in the Indian army and was also spending the night at Amrit's guest house, helped Amrit up the steps to look at his head more closely. The cut was deep, and bits of white matter were mixed with the blood. Sahila bandaged the wound and told Amrit that in the morning he should go to the hospital for stitches.

We were all shaken. The large empty room made it worse. I wondered what it was like for Amrit or for his wife, who had never wanted to leave Hedangna. We talked with Amrit to make sure that he really was all right. He told us that he had had a dream that a dog was trying to steal the last of the pig meat. He had leaped out of bed to chase away the thief. He was not used to the house, was not used to electric lights, and in his confusion had tumbled down the stairs.

Devimaya and I finally went back to our bed. Neither of us could sleep, so we turned on the fluorescent light and talked. We talked about choices and the future—hers, mine, Amrit's—and about the risks of leaving a place like Hedangna to face that future alone. I described the feeling of warmth, the feeling of rightness in Amrit's house the evening before he left the village and how different it felt here, in this tea shop in Khandbari, where we seemed to be working too hard to make a situation that seemed all wrong appear to be all right.

Devimaya was concerned about Amrit's head and wondered if he would be okay, but never having lived anywhere other than Hedangna, she did not comprehend how economic security does not necessarily lead to emotional security. She could not understand how people in the United States could live in big houses, with hot running water and glass windows, and still be unhappy, or that in leaving Hedangna, Amrit was leaving something that was not easily found in the world he had entered.

Devimaya eventually fell asleep while I lay awake on the hard wooden bed, staring at the magazine pictures of Hindi film stars taped to the walls, thinking about Amrit, thinking about the differences between Hedangna and the West. I remembered the farewell party that Amrit, Ganesh, and Raj Kumar had hosted the night before Brian had headed home six weeks earlier. They brought *raksi* and *jad* and a chicken to cook. Raj made a special trip to Gadi to get spices and salty snacks. We gathered on the second-floor porch of Ganesh's house, so we would not be disturbed. Amrit was in charge of the cooking. He wrapped a shawl around his head like a turban, so we would know that he was the chef, and he sat cross-legged on the floor preparing the food. We drank *raksi* while we ate the snacks and drank more *raksi* while we waited for the chicken and rice to cook. Ganesh started to dance; maybe one of us was singing. Brian joined Ganesh, and so did Raj Kumar. I think I was the one who was singing. We drank more *raksi*. And then Amrit made everyone sit down and listen while he gave a farewell speech to Brian. He stopped every few minutes and told me to translate. I can no longer remember a word of what he said. I can only remember Amrit being very serious, very sincere, and very drunk, doing what he had always wanted to do: preparing food and being a gracious host for Westerners, entering our world—if only for a night.

The next day, Amrit refused to go to the hospital. He was afraid of what people would say.

16

BAISETI THUMA

March 1993
SEATTLE, WASHINGTON

I stared at the list of the six kinds of coffee on the blackboard; the four kinds of muffins on the counter; the blueberry scones and oat scones; the chocolate chunk, molasses, and peanut butter cookies. People waited impatiently behind me, some even cut in front because I took so long to decide, and even then I was not sure, hoping the person behind the counter would assure me that I had made the right choice.

I walked to Green Lake, past one-story houses with short clipped lawns and cement sidewalks leading to screened doors. Someone drove up in a large car. A girl with long dark hair and tortoiseshell glasses slung her bag over her shoulder, said that she would see them tomorrow, and slammed the car door. This was normal, I told myself. Nothing was wrong. This was America, my home. There was no reason for my despair, no reason whatsoever.

I felt more disoriented than I had ever felt in Hedangna—lost and lonesome. I was lonely there and homesick, of course, but I expected to feel that way. And I knew that was where I wanted to be, no matter how hard it was. And now that I had left, now that I was back in my homeland, all I wanted was to go back there, to smell wood smoke on my clothes, to feel Baiseti Thuma's tiny, wiry hands gripping mine as she said hello. I longed for the wind and the rain and the wet, for the roar of the river and the taste of plain rice harvested from fields down the hill. I wanted the palpability of time, to be in time, not caged by lines telling me where I should be and what I should do and who I should be when I am there.

"What do you do?" I remember someone once asked William Carlos Williams. "I listen to the water falling," he answered. "That is my only occupation." But that was not true. He was a doctor—he healed people with his knowledge. He was a poet—he healed people with his words. He moved between worlds. I respond only to the world that I am in. When the place is open and inviting, I am open and inviting. And when it feels closed and cold, I become closed and cold.

◊

Two days later, Madeleine and I walked on the beach, rolling waves crashing onto the sand, the smell of salt and evergreens in the air. The silence of fog. Bald eagle overhead. Madeleine danced on the sand, the light returning to her eyes, the color to her cheeks as we had left her house in Seattle and driven to the coast, to this beach, drawn by a memory of wholeness, a longing in our bodies for food and water that is filled with light. There is nothing— no thing—that is more important, that is more crucial to our lives, to my life, to the life of us all on this planet. Not to follow this longing, not just to lose it—we all lose the thread, again and again we lose it, or else we would not be human—but not to follow it, not to look, not to smell, not to go to the source of your life, of my life, to the source of us all, is to die, to die of thirst, to die of hunger. To die because the food we eat and the water we drink does not have *charawa*—does not have the essence in grain that the ancestors bring, the essence that makes it last, the essence that makes us last, the thing I never knew was there. All I knew was the hunger, leading me along.

◊

March 1993
CAMBRIDGE, MASSACHUSETTS

I returned to Cambridge a month after the night in Amrit's tea shop. Brian had found a desktop and file cabinets and had created an office for me in the back room of our third-floor apartment. He had bought me a new computer and set it up on the desk, ready for me to use. He had hoisted large wooden lattices with a rope to the back porch because I had said that it would be nice to grow morning glories there, in order to create some privacy and soften the view of the cement garage and the asphalt driveway. Brian had waited for months for me to return, and he did everything he could to make me comfortable, to help me be happy, so that I would feel at home.

When healers return from their journeys into the realm of the ancestors, they call their *lawas,* call each by name, to make sure that none get left behind. I studied this as a scholar, not realizing at the time I left that I, too, should call my souls by name, call each gently and firmly to make sure that no parts of myself got left behind.

But I had no guide to lead me across the threshold of my return, to help me understand what I had learned and how I had changed, no one to teach me to incorporate those changes into the life I had left. And so I lost my way, not even knowing that my souls were lost until, years later, I found one of them sitting alone on a rock in Num, looking across the valley at the collection of stone-and-mud houses of Hedangna, which I had just left. An old woman bent under a load of wood was passing by. "Oh, *bahini,* oh, younger sister," she called out to me. "Why are you so sad?"

I blamed the distance and isolation on the culture and world I had come home to, not on the selves I expressed—or did not express—in that world. I could not explain this to Brian, was unable even to say what it was, exactly, that I missed. It was not what I had been studying in Hednagna; it was how I had felt while I was there. And so I told him that it was not the kind of computer I wanted, not because it was true, but because I did not have the words to express what was true, even if I had known what it was.

◊

Years later, it is still hard to talk about my return from Nepal and my reunion with Brian. Hard not because we did not love each other—we did. Hard because we missed each other. We missed each other when I was away, and we missed each other when I returned. Even now, when I am able to create the intimacy and openness I longed for but could never sustain with Brian, part of me still, probably always, misses the intimacy that comes from the rhythm and security of a shared history with someone who knows the taste of warmed *jad* or the look in Chute Rai's eyes. I am still surprised by the sadness of a loss, even as I understand the reasons for that loss. Part of me wishes that things could have been different. Looking for the foundation of the wall that was built, I wonder if it is the times we rushed over the thin places that instead called us to linger, inviting me to trust Brian enough to let him in, asking Brian to listen carefully enough to be deemed worthy of being invited. I am still sorry for a failure that had as much to do with anything either of us did or did not do, as it did with an absence of guides and a culture that has a lot to say about the values of following your "bliss" and moving on or of being responsible and staying behind, yet offers very little insight into what it takes to move, but to move in place.

◊

Looking back, I imagine that I did what others have done after returning from a long stay in rural Nepal. I marveled at the taste of good, strong coffee; the feel of the thin stem of a wine glass between my fingers; and the colors of food on my plate: green and yellow and red, all in one meal. We spent time at Brian's parents' second home in Canaan, New Hampshire, where we did things that brought us closer together: chopped wood, cleared brush, skied and biked, and went on long runs on trails through the woods. Brian bought sheep. I planted a garden.

That year, Brian became involved in an effort to create the Good Life Center, an organization to carry on the legacy of Helen and Scott Nearing after Helen's sudden death in a car accident. He also began to work with a homesteader and craftsman in Down East Maine, helping him protect the land around his home. I resonated with the ideas at the core of these projects; I, too, wanted to nurture and support ways of living that were simpler, healthier, and more intentional and was grateful for the opportunity to interact with others who shared these goals.

Yet I increasingly felt distanced by the process of achieving them. The work absorbed a tremendous amount of Brian's time and energy, but that was not it. Even though the individuals I met in these circles had interesting ideas that were compelling and important, interactions with them often left me feeling empty, the invisibility, perhaps, that arises around people who are so clear about their own views and vision that they have no room to consider anything else. I thought of the generosity of spirit I had experienced with Chute Rai and Baiseti Thuma, their humility and my feelings after interactions with them compared with those after meeting with these people, whose values I shared.

In the early summer, Brian and I attended a retreat held after Helen Nearing's death. Friends and others connected with the Nearings' life and work gathered to discuss the future of their home, Forest Farm, and their legacy. There was much talk that weekend about how powerful, important, and inspirational the Nearings had been. There was also much talk, outside the formal gatherings, about the gap between the image the Nearings had projected of their life and the reality of that life, as described by the neighbors who knew them best.

I found that I was less interested in the details of this gap than in the reasons for it, wondering why the Nearings had told the stories they did and not others, curious about the larger context from which their stories had emerged. If what the neighbors said was true, I wondered how the Nearings had felt about living a life that was perhaps more filled with compromise than they cared to admit, especially when much of their reputation was based on their unwillingness to compromise. I could not help feeling that there might be more to learn from the gaps and silences in the story, from what the Nearings had chosen not to tell, not just what they had.

A month or so later, I spent a week house-sitting at Forest Farm. I intended to use the silence and the solitude to go more deeply into my own work, but Helen and Scott's presence was too powerful. The house and the land seemed to be an extension of them, an expression of their lives in physical form. This presence demanded a response, as a living person would. And so I shut off my computer and spent the week working in the garden, looking through the books in their library, watching the sun rise across the ocean, and giving tours to visitors who had driven from Alabama and North Carolina—absorbing what was in the place rather than reshaping it to fit my needs.

At the end of the week, driving back to New Hampshire on a hot summer day, I got stuck in a line of traffic on a back road in western Maine. In front of me was a truckload of dead chickens; around me, the road was being paved. I sat in my pickup truck, white feathers swirling against the windshield and tar fumes wafting through the open windows and had the visceral sense that these were the forces that Scott and Helen had fought—the institutions that created and sustained a world filled with factory-raised chickens and toxic fumes. In that instant, I understood why the Nearings had devoted so much energy to upholding a particular image for the world, even if they had not always lived up to that image. That is what it took to pave a pathway through the poison.

And so I dismissed my feelings, assuming that there was something I was not seeing, just as I dismissed my unhappiness as a graduate student at Harvard and my increasing loneliness in my marriage.

◊

During the day, while Brian was at work in downtown Boston, I began to read through my handwritten journal notes. I skimmed through pages about which crops were grown and traded, through endless details about different land disputes. I found lists of flower names and plant names, sketches of altars built for the different ancestors, and reflections on anthropology and international development. I flipped through the pages until I came upon a detail that suddenly took me back to Ganesh's darkened kitchen. I had just hauled a load of rice up the hill from the fields. The rice was heavy, the work was hard, and it was always nice to linger by the fire, drinking *jad* and eating chili peppers mixed with salt. A steady line of people carrying baskets filled with the year's harvest climbed the ladder to the storage bins upstairs, and then joined us by the fire. Jaisita refilled our bowls with *jad*. Everyone was slightly drunk, joking, speaking about things that were not important, talking to hold onto the moment when their bellies were full, the weather was good, and there was plenty of rice in the grain bins.

And then I was at my desk, looking through the window at an abandoned lot surrounded by a chain-link fence. I continued to read my notes, searching through details of what had been said and done, the information I needed for a dissertation, the information that had taken me to Hedangna, turning pages until suddenly I saw Chute's face across the fire. He was speaking to me in Nepali over a group of loud men who were drinking *jad* and speaking in Yamphu, which I could not understand, his voice like a hand reaching out in welcome. Or I felt Devimaya's stubby, rough hand clutching mine as she said through her sobs that she had never cared for anyone the way she cared for me, gripping my hand as we said goodbye.

This was what I wanted—more moments, more memories, more sounds and smells, more gestures, more details from the moments when a connection was made. I wanted, as T. S. Eliot writes, the "quick now, here now, always" (1971:59). Not all that stretched before and after.

But the before and after were what made the moments possible.

◊

Each morning, I went to my office in the back room to sit alone before my computer all day and try to write a dissertation. I talked to the clerk at the grocery store, the two friends whom I called several times each day, and Brian when he came home in the evening. If a villager was pulled out of Hedangna, a whole network of cousins and sisters and uncles and grandmothers would come tumbling along. If I were pulled out of Cambridge, one or two people would come with me, but the thread would be thin and could easily break.

◊

One night, I told Brian that I needed more intimacy, that I had to share more than ideas, that I needed him to want to be closer as well. I said that something inside was disappearing, something that had come alive in Nepal. Brian replied that he did not feel the same need for closeness, the longing for connection. What he needed was a structure he could count on. He had to know that he could trust me to respond rationally to things that he said or that I experienced. Most people were happy, and he did not understand why I was not. I had so much. He felt enormous pressure at work, he said, and wanted consistency and dependability at home, not another person who needed something more.

Another time, we sat in the field above his parents' home in Canaan. We had spent the day clearing the land, and we felt close and connected. We watched the sliver of a moon rising in the sky, and Brian told me that he was a catcher, a catcher waiting to hold me when I decided to come home to stay. I had said that all I wanted was to be held, and he responded that I would not let myself be held. I was quiet, because I knew it was true.

◊

A year after returning to Cambridge, I presented a paper in the seminar room for the Department of Anthropology at Harvard. The room is formal: a long heavy wooden table, wood-paneled walls, an Oriental carpet, stiff-backed chairs, and tall windows that let in light but not air. People spoke in hushed tones in that room, and laughter always seemed somewhat forced. The weekly seminar followed an unspoken protocol: after a paper was read, tenured faculty asked the first questions, followed by assistant professors. Graduate students, especially those who had not yet conducted their research, rarely spoke.

I sat at one end of the table; my adviser—a tall, regal woman with silver hair—sat at my side. I had been anxious about this talk for months. But I had written it out and thought that the only thing I had to worry about was to keep my voice from shaking.

I began to read, and my nervousness subsided. I was even able to be slightly animated and make people laugh. I realized that I probably would do fine and began to relax. But then suddenly, around page 5 or 6 of a twenty-three-page paper, I began to sense the presence of Baiseti Thuma above my left shoulder, near the top of the windows outside, looking in. I could not think about this grandmother without an ache in my chest, an instinctive turning away because I did not know how to handle the depth of my feelings, did not know how to express or share them in a way that was not trivial. And here she was, showing up in a room and in a moment where she could only cause trouble.

I looked around the table at the nodding heads and knew in an instant that I had gotten the whole thing wrong, even as I thought I had gotten it right. I was telling a story—I thought that was enough, that made my paper different from other scholarly papers. But it was my story, the information from the villagers lives filtered through the perspective I had learned about what information mattered and what did not. There was no room in my words for Baiseti Thuma's life, no room for the expression on her face—wide-mouthed, wide-eyed—as she gripped my hands and taught me how to dance, no room for the anger and loss she had suffered from the land dispute about which I now talked, no room for her story even though that was precisely the story I had wanted to tell. I was talking about lines without recounting what went on inside those lines, inside the boundaries of the villagers' lands, inside the borders of our skin.

I thought of Chute Rai's comment, my last afternoon in Hedangna, that everything I needed was in the song. The meaning, I now realized, was in the song—in the song itself—not in the meaning I made of what was there. As I had learned about land tenure, political disputes, rituals, and farming practices, as I began to understand the circumstances of the villagers' lives, the distance between the villagers and me decreased. Yet that knowledge kept the boundary between us intact. I was in Hedangna for a reason: to discover underlying patterns and themes that would demonstrate the anthropological significance of my work. Rather than a path out of William Cronon's black box, my research kept me trapped inside. Even as it allowed me to enlarge that box, as long as I was gathering this information for an academic audience at home, I interpreted the villagers' world through my own cultural framework, which determined what counted as knowledge and what did not. I was turning their experiences—their lives—into what on some level ultimately destroyed the wilderness in "The Bear," turning those experiences into data I could use, information for my benefit, not theirs.

I thought again of what Cronon had told me during our twenty-minute meeting. I wondered what it would take to let myself be transformed by what I had encountered, not just defend—like a plot of land—what I already knew. What shifts did I still have to make to see that the people I had met and the places I had visited—the experiences I had had—were like Minaba and Sepa's wooden bowl and walking stick, guides leading me down the river, to a world I could not yet see?

◊

Shortly after my presentation at Harvard, I attended Watershed, a gathering in Washington, D.C., of poets and writers to talk about nature, community, and a sense of place. I sat in rooms filled to capacity and listened to Barry Lopez, Richard Nelson, Terry Tempest Williams, Wendell Berry, Rachel Bagby, and others talk about how the economy has created a pain we cannot live with. They said that the economy was breaking the rules and breaking our hearts and that we needed our imaginations to get ourselves out of the mess we were in. Imagination, Barry Lopez asserted, has the power to transform what is closed and about control into something open and alive.

In the evenings, there were more readings in a church on the west side of the Library of Congress. Anthropologist and nature writer Richard Nelson quoted Gary Snyder about the secret and the secret within. And later Linda Hogan, a Native American poet and writer, read a poem about going into something unknown, something deep and old where there is no boundary, no division, between this world and the other world.

Toward the end of Linda's talk, I suddenly sensed the presence of Baiseti Thuma far up on the left side of the church, just inside the stained-glass windows. She had slipped in quietly, unnoticed. Certainly I had thought of her since the day of my presentation at Harvard, but I had called her up. And now here she was again, coming on her own into my world. But this time, in this room filled with care and compassion—where the language I hungered for was spoken openly, with grace, and where people were not afraid to say what they loved, to love what they loved even as they saw it being destroyed, to not run from the pain of that destruction—in this room, she stayed.

◊

My last term at Harvard, I took a nonfiction-writing class. The first day, the professor told us that our words could never capture what we hoped to say— life was too complex and mysterious—but we still should try, since trying was the most important thing that we could do as writers. This was the one lesson I could say with confidence that I had learned in Nepal. Yet I had never heard another professor say anything similar in any other class, had never heard it expressed in any of the circles I moved through with Brian. Everyone else had answers. I just had to read more, do more, talk more, and I would find the an- swers as well. There was something I was not seeing or something wrong with what I saw. This was the first time I heard someone suggest that what I did see was true.

◊

I dreamed of a child reaching for my hand. In the dream, I am both the adult and the child. Snow falls silently outside. It is Christmas Day. The presents have been opened. No one is around except the child. Pulling on my hand, she takes me to the far side of the Christmas tree. We kneel down. She hands me a gift that she has opened.

"This was meant for you," she tells me.

"No, no," I say. "It is for you. I think it is meant for you."

"But it isn't what I wanted," she replies. "I didn't want a thing. I wanted green fields and flowers and a stream behind a house. I wanted a clearing in the woods. I wanted the sun."

"But they didn't know," I say, thinking of my parents, thinking of my daugh- ter—thinking of my longing as a mother to say the right thing, to give the right gift, to make my daughter happy. "They thought this was what you wanted, that this would make you happy. They didn't mean to make you sad."

She nods.

"I'm telling you what I do want," she says, "so you know. Now go and find it."

17

A FAR-OFF PLACE

September 1995

There was now an international telephone line in Tumlingtar and fifty-eight national phone lines in Khandbari, a walk of two and a half hours up the hill. There had not been any when I had been here before. At least seven foreigners lived in Khandbari; two years earlier, there had been only one, a Peace Corps volunteer who was rarely in the village. The Makalu-Barun Conservation Project had finally opened its field office in a three-story building above the dusty site of the weekly outdoor market.

I went to find Amrit the morning after my arrival. He was wearing the same blue sweatpants he had been wearing when I had seen him last. He smelled of old cigarettes. He had moved his family across the dirt path from his first tea shop and into one of the few remaining one-story, thatched-roof mud buildings, where the rent was cheaper. He and his wife had had another son. They paid the rent with money made selling *raksi* and single cigarettes. There was not much for Amrit to do in Khandbari, so he agreed to help carry my bags to

Hedangna. I asked him so we could have a chance to catch up and so I could help him out by giving him money.

As we headed out of Khandbari the next morning, Amrit said that his life there had turned out exactly the opposite of what he had expected. He thought he would make friends quickly. He thought he would be able to do business. He thought he would earn money. He had stayed in the building where I had last seen him for three months. The rent was 1,000 rupees a month, which the family could not afford, and finally his wife insisted that they move to a less expensive house. The new building was not as good, but the rent was only 300 rupees a month. Amrit had just heard that it had been sold, so he was likely to have to move his family again.

He told me that his wife wanted to return to Hedangna, but the mud on their house there had become dry and cracked, and the roof was starting to collapse. Then he said that he planned to move back to Hedangna the following autumn, in time to cut and store firewood for the year. And a bit later, he told me that he had mortgaged his house and land to another villager to pay back debt incurred since moving to Khandbari. He could not return to Hedangna until he had enough money to reclaim his fields. Plus, he added, back in Hedangna, he would have to do hard, physical work.

He was quiet for a while as we made our way up the long hill out of Khandbari, and then said that he really planned to move even farther south to set up a tea shop along the road near Hille. There would be more business there. Down there, he said, he would finally make some money.

Amrit had always had a greedy side, but I enjoyed the curiosity and boldness that made him seek me out, the same curiosity and boldness that made him look elsewhere for a different sort of life. When I had known him in Hedangna, the rhythm of life—the cycle of chores, the rituals, the ties of kin and neighbors—had provided a structure to contain the longing he felt for life away from the village.

In Khandbari, his wife prepared the food, washed the dishes, and hauled the water and wood. There was little left for him to do: no fields to plow, no livestock to tend. His neighbors were not his relatives and did not share his language and traditions. He planted no crops, so there was no need to perform the rituals to call for rain or keep away the hail. There was no reason to keep a *mangsuk* (shrine to the ancestors) because the house he lived in was not his own. "Why honor someone else's ancestors?" he asked. When he or anyone else in his family fell sick, they walked to the hospital on the other side of the bazaar instead of calling on a shaman or priest.

While staying in Khandbari, I would often see Amrit walking up and down the dirt path through the bazaar. We would stop to talk. He asked what I was doing, where I was going. When I asked him, he said that he was simply walking up and down the path. Sometimes I passed him on the edge of a card game looking over the card players' shoulders, because he had no money to play himself, or talking and drinking *raksi* with villagers from Hedangna who were obliged to stay in his tea shop, passing the hours until it was time to eat or time to sleep. Without any work or any village rituals and responsibilities to center him, Amrit seemed to be propelled by nothing but his desire and his discontent, his life slipping away as he walked up and down the path, waiting for something to happen.

The trip from Khandbari was slow because neither of us was feeling well and because Amrit had lost the habit of walking long distances over rough terrain. His pace slowed to a crawl on every steep incline. As I listened to him, I found myself getting more and more impatient with the choices he had made and the desperation these choices revealed. His longing and dissatisfaction depressed me less because of his particular situation than because I did not see an end to the conditions that had made him want to leave Hedangna. Now that he had started moving, it seemed there was nothing that would enable him to stop anywhere long enough to feel at home. It was less that he had failed than that his "culture"—that inherited from his forebears and that passed on in schools—had failed to provide what he needed to make his way in a world that was neither that of his ancestors nor that he had imagined while living in Hedangna, listening to radio advertisements and travelers' tales.

The next morning, Amrit and I finally reached the Arun River, just below Hedangna. As we climbed the steep hill to the village, I noticed—for the first time since setting out on the journey—the ground under my feet, conscious of each step on the packed dirt and stones on the trail to Gadi. I thought about how long it had been since I had been aware of my feet on the earth, how long it had been since I had heard the sound of nothing but river and wind. At home, I experienced this heightened sense of perception by going away into wilderness, away from people. I kept returning to Nepal because here, for me, the boundaries between people and wilderness were blurred, a blurring that made it possible to encounter the sacred in the everyday, not only in the wild. I forgot that coming to Nepal was like being on a pilgrimage and that, as much as being in Hedangna, made the sacred so accessible, more accessible than it was for anyone—American or Yamphu—to experience at home.

We climbed the hill past fields with ripening rice, clusters of stone-and-mud houses, pigpens and chicken coops, sheds for storing wood, and narrow corrals for confining livestock at night. Scrawny chickens pecked at grain along the edges of the packed-mud courtyards, and young children chased one another along the network of dirt paths winding between the houses. We walked past these houses and through the degraded forest, and thirty minutes later entered the mud courtyard of Dilli Prasad's house. Dilli and Dhanmaya saw us arrive and came down from the millet fields above their house. Baiseti Thuma, wrinkled and thin, looked up from cutting millet, greeted me with a wide toothless grin, said that she would come by later, and returned to her work. I had not seen her for two years. She looked tired.

Amrit headed down to Ganesh's house. I took off my pack. Dhanmaya went into her house to make rice beer. We ate, talked about the rice harvest, and caught up on news about Brian and Devimaya. They returned to their work. I went to wash in the one small semiprivate water hole on the Naong River. It was hot, sunny. I was back.

◊

To return to Hedangna after two years in the United States and feel as though I had never left was unsettling. There was not a lot to say: just planting and digging and carrying and cooking and eating and watching and waiting, living the cycle around and around until, one day, you die. My same anxieties about research returned. When I was here before, this cycle, this sameness, had been a source of peace and security. This time, I felt a kind of heaviness, a sense of stasis that was almost oppressive, which I had forgotten about while missing Hedangna from Cambridge. Places I had visited in the past two years and people I had met felt distant, unreal. What mattered was here: the hunger in my belly, the sun drying my hair, the wood smoke drawing tears from my eyes.

For two years, I had longed to return to Hedangna, hoping to find a thread to connect me with the time that was fieldwork. Now that I was here, I remembered all I had forgotten: the heaviness of time, the aloneness, the tasteless food, the hunger, the sameness of the work stretching from one day to the next. I could not just wake up and walk smoothly through my day, as I had in my memories. When I translated texts, I worried about whether I should instead be talking to Chakra Bahadur or Suresh or someone I had not even considered. When I spoke with the men, I wondered if I should be talking to the women or not talking to anyone and instead just helping them in their fields, observing and absorbing. I could not find my bearings in this place, where little was readily available to mark the movement of time except more babies and fewer older men. Now that I had finally arrived, I realized, like Amrit, that I just wanted to keep moving.

◊

A few weeks later. I was writing in the glow of an electric light, alone in a second-floor room of a four-story hotel in Thamel, on the north side of Kathmandu. I had just finished eating a grilled cheese sandwich. This morning, I had been in Chainpur, a two-day walk from Hedangna, on my way to spend another day talking to court clerks about a land dispute I had spent months gathering information about in Hedangna. But then I drank another cup of tea, went to my tiny room to get my notebook, and instead packed my belongings and put on my backpack. I headed toward the courthouse but then kept walking, past the courthouse; away from the idleness of government office space; from peering eyes; from comments about my height, my clothes, my solitude, my load, my skin color; from another plate of watery dal and bitter greens in a windowless room with ten male government workers. My feet kept walking, and the next thing I knew I was halfway to Tumlingtar and realized that there was a chance I could get a flight to Kathmandu that day. And I did. And here I was. Even as I walked more quickly to catch the plane, I kept telling myself that I should not leave yet; I had more questions to ask. I needed more information about the government perspective on disputes and land rights in Hedangna, without which my research would be incomplete. I kept telling myself to stop and turn back, to do it the right way. But my feet kept walking.

The next night, I met Robert, a friend whom I knew from my previous work in Nepal. After dinner on Durbar Marg, we lingered, talking about headmen and villagers and the journeys of the shamans, about what one saw when one moved between worlds. After our meal, we walked through the slowly emptying streets of Kathmandu. We walked for a long time, still talking. I had not spoken this much for so long, had not spoken—ever—with anyone who understood so completely the ambivalence and the allure of fieldwork, who also cared less about the research or the theory than about the opportunity it offered to experience life fully, nerve endings bare—now, while there was time. Someone who also felt more comfortable in this place, where you never knew quite what to expect next, who preferred moving between worlds to settling into any particular world.

It was late. It was time to head toward my hotel, but I kept walking. It was so comforting, this talking, this walking. A lonesomeness I had felt since leaving Hedangna, since returning to Hedangna, felt less acute, as though this walking could help me find my way to a home I had never known. Walking, walking. Watching, watching. Robert finally stopped by a taxi. He was staying on the other side of town. My hotel was close by. We kept talking. I stood by the open taxi door, knowing that I should turn away, knowing that I should walk alone through the dark streets and back to my room. I knew that everything that followed depended on what I chose. I knew what I should do, knew that I should stay with who I knew myself to be, with the self I had promised others I would be. And yet I also knew that, on some level, I was simply following the thread that had led me away—walking, walking, watching, watching—leading me deeper into the forest I had already entered. I ducked my head and climbed into the cab.

◊

A few days later, I sat at a small desk. A large mirror was on the wall behind my computer. A pot of cold coffee was at my feet, notebooks and papers spread out on the floor. I was trying to write about sheep and shepherds, about grazing rights in the high pastures, which was really about how the stories of those with less power were getting subsumed by those with more. I stared at the screen, trying to make myself think about sheep and yaks.

I listened instead for the telephone. I listened for motorcycles on the dirt road outside. I ordered more coffee. I went to the window and stared at the rickshaws and taxis crowding the dusty side road. I sat back down in my chair and looked through my notes, trying to remember what I had learned about grazing rights. I wrote a sentence. I went back to stare out the window.

There was a knock at the door. I opened it. He entered. I shut the door. He took off his jacket and put down his bag.

Being with me, he once said, was a way of encountering himself; being with him was a way of encountering my self, a self I had never known. The self that came alive by going on my own into the wind and the rain and the wet, that came alive by seeing threatened all the things I held most dear. I had known as I climbed into the cab, knew again each time I opened the door, that I was risking everything I had built my life around. Yet the risk of not opening the door seemed greater.

We made love. He put on his jacket and picked up his bag. He said that he would call. He opened the door. He walked out. I shut the door. I sat back at the desk. I turned off the computer and stared at my reflection in the mirror.

◊

Later that week, I contacted villagers from Hedangna who had moved to
Kathmandu. I met Ganga on a clear, cool September morning outside the
Bhat Bhatini store on the northeastern side of Kathmandu. Ganga was one of
a handful of young men from Hedangna who had come to Kathmandu for further studies, following in Raj Kumar's footsteps. He greeted me as I climbed
off my bicycle. Ganga had been in Kathmandu for more than a year. I had not
seen him since my last trip to the village two years earlier, and I was curious
to hear how he had been. I pushed my bike as we walked along the edge of the
pavement toward the outskirts of town. Black tempos (three-wheeled taxis),
spewing dark clouds of exhaust, sputtered past. Rickety rickshaws rattled
over the potholed road.

We walked down a hill and across a newly constructed bridge. Cattle and
water buffalo grazed along the banks of the Dhobi River and waded in the
shallow water. Women washed sheets, spreading them to dry amid the garbage strewn along the banks. Two- or three-story, cheap reinforced-concrete
buildings had been constructed hastily to accommodate Kathmandu's growing population, extending the boundaries of urban sprawl into former rice
fields. Cement blocks marked future construction sites on empty lots. Paper
and plastic bags littered the sides of the road, and mangy dogs sniffed through
the garbage. Nothing green or growing was in sight.

I followed Ganga into a bare courtyard, where men, women, and children
were sitting and talking or washing dishes on the cement walkway, and into a
room on the first floor of the apartment building. Inside the eight- by ten-foot
room, his two roommates squatted on bamboo stools, busily chopping potatoes and garlic. In the corner, beneath the one window in the room, were two
kerosene stoves, several pots, and a bamboo basket that contained dried chili
peppers and two or three red onions. The hissing of a pressure cooker and the
fumes from the stoves filled the small space. There was one bed in the room, a
bamboo bookshelf, and blankets and clothes stacked neatly against the wall.

I sat cross-legged on one end of the bed. I had just returned from Hedangna, so the men asked a few questions about what people were doing back
home, whether they had started planting millet, and who had been married.
Then they talked about Kathmandu, about school, and about their days spent
wandering around the city ever since classes had been cancelled because of
the political strikes in the country.

As they talked, I wondered how they felt to be so close to this world filled with Land Cruisers and computers, with VCRs and desk jobs, a world that they longed to enter but that, now they could finally touch it, seemed to recede even farther from their grasp. They had electricity, a cement floor, and a tin roof: all the things many villagers said they longed for in Hedangna. They cooked on a stove and got clean water from a pipe outside. But this room was no place, or it was any place. They were three more anonymous students living in a cheap, cold cement room on the margins of the modern world. Despite their physical distance from Hedangna, a future plowing their fathers' fields back home— a future they kept referring to as the mark of their failure—was more concrete than any they could imagine for themselves in Kathmandu.

◊

When I first went to Hedangna, Raj Kumar did the work he had to do—plowed his father's fields and hauled loads of rice on his back—but he did it reluctantly. There was a distance between Raj and those who lived in the village, in part because there were things he simply did not know how to do: weave bamboo baskets, construct bamboo fences, or carve wooden plows. He did not know—or want to know—much about the past. But the distance was deeper than that. He obviously cared for and respected his family—he had returned home because they asked him to—but he looked with disdain on their long-standing farming practices and dismissed the stories passed on by older men, saying that they were not in the history books he had read in school, so they must not be true.

When I passed through Khandbari on my last visit to Hedangna in 1997, Raj had taken a job as assistant to Khagendra, the community-development officer at the Makalu-Barun Conservation Project, and he had moved his family to a two-room apartment just north of the bazaar. Each morning, after eating a plate of rice, dal, and vegetables, he walked through the bazaar to the three-story wooden building that was the project headquarters. He had his own desk, his own responsibilities—he had a place in the larger world he had always longed to enter. And yet he still wasn't satisfied. The work was boring, he complained, except when he was out in the field meeting with villagers. He complained about the job, which wasn't permanent. He complained about having to return to Hedangna, which he predicted he would have to do in a year or two. His position was more secure than Amrit's, yet their discontent seemed the same.

Several months after leaving Hedangna in 1995, I had received a letter from Devimaya written from Uttarpani, where she was studying to become a veterinary assistant. It was a short letter, written in Nepali. She described Nepal as a poor and backward country whose people were illiterate. They needed development so they could become modern, like people in the rest of the world. I had never heard her say anything like this before. It was similar to much of what I had heard from Raj Kumar, more like a passage from a school textbook than what Devimaya actually believed. I wondered where it came from.

Instead of building on the traditions and knowledge of the villagers in Nepal, the education they received taught them that the places they came from were ignorant and dirty. Their culture, their religious beliefs, and their ways of using the land apparently were what prevented them from entering the bright, shiny modern world, not the reality that in that world there was not enough—enough opportunities, enough resources—to go around. They were taught that their poverty was the result of who they were and where they came from—things they could not change—not the result of the deep inequities in the world they hoped to join.

◊

Men and women have always left Hedangna. Village headmen went to the district center and even to the national court to dispute land claims and deliver taxes. Priests and shamans traveled into the invisible world of the ancestors. Men and women left to work in northern India. Some stayed. Those who returned to Hedangna after several years told me that they had come back because their land was there and they had to care for their parents. They compared their travels with my own, saying that I was living in their village for a while, but it was not my home and it never would be. They had returned to Hedangna, they told me, because it was theirs.

These individuals left on a path well worn by their predecessors. Younger villagers were part of a new generation of Yamphu, a new generation of Nepalese, who were leaving villages by a different road, a road that might bring them home physically but that risked leaving them stranded, intellectually and emotionally. They joined the growing number of younger men and women who had experienced enough of the modern world to be dissatisfied with the life of their parents but who still lacked the education, money, or connections needed to cross into a life other than what Ganga and his friends had found: a life that was shabby and desolate, on the margins of what they hoped for.

When Tibetans left their country, they had a written language, an established religious tradition, and an articulate leader to keep their culture alive.

Although of a place, the Tibetan Plateau, their stories and beliefs could travel. The oral tradition of the Yamphu is woven into the landscape that is the home of their ancestors. The stories told and passed on in Hedangna help the villagers live in a place much like it was a hundred years ago, a place with no lights and no roads. When they leave this place, they leave more than the physical landscape where they were born. They leave their emotional and cultural home as well.

According to the tales of Shambhala, the danger in setting out in search of a hidden valley is not that you might die along the way—that danger exists even if you stay at home—but that you lose your way: that you leave your home and never reach your destination or that you are unable to recognize it when you arrived, so you keep searching and searching for a place that fits an image in your mind. Your soul becomes lost, not in the past or even in the present, but in an idea—a dream—of a future.

On their journeys into the world of the ancestors, into the past, shamans and priests in Hedangna are skilled at retrieving souls that become lost. They have less experience with finding souls that become lost on journeys into a world that, for most Yamphu, is even more inaccessible than the spiritual realm. The journeys of the priests and shamans into the world of the ancestors are journeys of remembrance: the healers remember what has become divided. The journeys of this new generation of men and women are journeys of forgetting.

I had gone to Hedangna because I believed that indigenous cultures had a wisdom about how to live on the earth that the industrialized West had forgotten. Yet, instead of the far-off place of my imagination, I found men and women who shared my restlessness, people also driven by the feeling that there must be more to life than they could find in their village. I went to a farming village a week's walk from a road because I wanted to get closer to the land, to other people, and to the sacred. The villagers who befriended me wanted just the opposite: distance from the demands of neighbors, from the whims of ancestors, and from the harshness of the land. What we believed we might find on our travels was shaped as much by myths—myths of the wisdom of indigenous peoples, myths of the ease and plenitude of the modern world—as by reality. We believed that something could be found in another place that could not be found at home. We blamed our dissatisfaction on something in the world, not in ourselves or in the stories we told ourselves about that world. If only we lived elsewhere, we would feel at home.

◊

I had stayed with Raj Kumar and his wife in Khandbari before heading to Hedangna with Amrit. Raj told me that Hedangna had received a small grant to "improve" the *tsawa* as part of the village cultural conservation program. But the money had gone to the wrong people, and the improvements had not been finished. He said that I would not like it.

A few days after I arrived in Hedangna, I visited the *tsawa*. Even though I was cynical about international development, even though I was disillusioned with the Makalu-Barun Conservation Project, I still was not prepared for what had been done to the *tsawa* in the name of preservation. A large cement holding tank had been built in the middle of the pool of water where the five stones still stood, to the side of the tank. Five smaller cement blocks had been built onto the front of the cement tank, I guessed in honor of the founding fathers of the village. Metal pipes for running water extended from each cement block. The stone spigots characteristic of most outdoor watering holes in urban and rural Nepal were gone. Graffiti had been carved into the side of the holding tank and scratched onto the wooden slats along the side of the tank. A soap wrapper floated in the pool of water beneath one of the pipes, and a black plastic boot had been tossed to the side.

The *devithan*, a shrine for the goddess of this place just above the *tsawa*, also had been "improved." Turquoise paint spilled onto the whitewashed building from a sloppy paint job, and the new stairs were crooked. It was not finished either, Ganesh later told me. The money was instead being used to build a school in another part of the village. Three teachers had been in charge of the project, and the money and responsibility for the improvements had gone to them, not the villagers. So the villagers did not really care what happened, Ganesh said. They invested nothing in the project, and they lost nothing.

The platform was still cool and quiet. It was the same peaceful, shady retreat, enlivened by the sounds of birds singing, water running, and children and crows calling from a distance. A woman who was getting water explained that it was better, because the water now flowed through metal pipes instead of stone.

◊

I was raised in West Virginia, where the mountains were being emptied of coal to fuel far-away companies and the air was filled with chemicals to make plastics for use by people in other places. I lived in a state that was on the dark side of "progress." So I imagined that other places, those that were on the bright side—or were beyond the reach of "progress"—must still be enchanted, even if the one where I lived was not. I believed that even if cement was covering the beautiful places in my own home, there was still some far-off land where the poison could not reach.

Maybe for the Yamphu, the water—the same water—tasted better after flowing through metal pipes instead of stone channels. It was not for me to decide. And even if this cultural preservation project was not an improvement, perhaps it was not a loss for them, or at least not the loss it was for me. I turned and walked back up the hill to Dhanmaya's house, not at all sure where to go from here.

18

ABSENCE

February 1997

My last trip to Hedangna was in February 1997. Raj Kumar was still in Khandbari, and I visited him in his office at the headquarters of the Makalu-Barun Conservation Project. He talked about his work, his wife and sons, and his family back in Hedangna. He filled me in on Deuman, his brother, still studying in Kathmandu. And just before I left, he told me about Dev Kumar, whom I knew had been in love with Devimaya since I had first lived in Hedangna six years earlier. Dev Kumar and Devimaya had planned to get married later that month, Raj told me. Then his voice changed, and he told me that three weeks ago, a month before their wedding, Dev Kumar had fallen off a cliff and died.

◊

When I arrived in Hedangna a few days later, I heard that a leather wallet had been found in the ashes after Dev Kumar's cremation, the only thing that remained after his body had burned. Inside the wallet was a 50-rupee note and a black-and-white photograph of Devimaya. Devimaya's aunt brought the photograph when she came to Hedangna several weeks later for Kumari's marriage to the man with whom she had gone to India the previous year and who was now the father of her child. On the morning after the ceremony, Devimaya and her aunt sat inside Dilli and Dhanmaya's house, watching the fire and preparing *raksi* from the *jad* left over from the wedding, alone for the first time since the celebration had begun. Her aunt handed Devimaya the photograph. She did not say anything at first and then asked Devimaya why she had given it to Dev Kumar.

Devimaya looked at her tightly clenched fists resting in her lap. Finally, she said, so quietly that her aunt had to lean forward to hear, "Because he asked." The aunt shook her head. She looked through the doorway at the darkening sky and then at her niece. "Why did you do this?" she asked. "Why did you fall in love? Why did you *let* yourself fall in love? Now, look what you have done. Look at this sadness, this grief, you have caused it all yourself."

◊

Dev Kumar had died on his return to Uwa, where his family lived, from the Barun festival in the first week of 1997. When he and his friends reached the stone platform at Pathibhara ridge, just above Uwa, they stopped to buy *raksi*. Dev Kumar had just completed his exams in Kathmandu; he had not seen his friends for a long time; and this was his last trip to the Barun *mela* as a single man. Perhaps he was celebrating. He drank two bottles of *raksi* and then set out ahead of his friends. A little later, he stopped on the side of the trail to take a nap. He awoke, startled, leaned forward too quickly, and tumbled over the cliff. His brothers found his body.

I had come to Hedangna for a three-week visit. I had not known that a wedding had been planned, but I had seen Dev Kumar's pictures, read parts of his letters, and known for years that he and Devimaya were in love. My first evening in Hedangna, as Devimaya and I spread our blankets by the fire to sleep, I said quietly that Raj Kumar had told me what had happened.

"I'm so sorry, so very, very sorry," I said.

She dropped her head for a moment, and then quickly brushed her hand over her eyes.

"It doesn't matter," she said. "It is nothing." She paused. "What is there to do now? He is dead. There is nothing more to say."

◊

Dev Kumar was the most highly educated Yamphu man in the area; Devimaya, the most highly educated woman. He was twenty-six years old, the youngest and cleverest son of Chakra Bahadur, one of the most politically astute head-men in Pathibhara District. She was twenty-four. They had been writing let-ters to each other for seven years; he had been asking her to marry him for the past three. The first time Dev Kumar proposed, Devimaya told him that she had to continue to study so she could get a job and provide money for her family. She would marry him in five years, she said. Each year he asked again, and she replied that there were still three years, or two years, to go. The past autumn, after she had again put him off, Dev Kumar refused to listen. "We're going to die soon," he insisted. "We have to get married. Now." She finally agreed, and they began to plan their wedding.

Devimaya had last seen Dev Kumar in December, when they spent the afternoon in Biratnagar, a bustling town in the eastern Terai. She was there to study for an exam, and he was on his way back to school in Kathmandu. They sat in a dark tea shop, eating goat meat, drinking *raksi*, and talking about where they would live and what they would do once they married. Then they went shopping for Indian cloth—he as a gift for her at their wedding; she for her sis-ter and, perhaps, herself, to wear at each of theirs. He said farewell and climbed aboard the night bus to Kathmandu. That was the last time she saw him.

◊

There was a wedding in Hedangna the day after I arrived. February, the month with the least amount of work, was the time of weddings, and this had been a particularly busy year. Devimaya had not attended any, except Kumari's, and only because she had no other choice. But this morning, Dhanmaya kept talk-ing about the wedding. She told Devimaya that her *mitini* (ritual sister) had arrived and that the two of us had to go together. Devimaya said nothing, but just went silently about her morning chores. Later, as her mother wrapped her shiny purple lungi with the bright-yellow dragon around her hips, she told me that Devimaya was not attending the wedding because she was too embar-rassed about what others would say and that I should go later with Kumari. She then headed off to the wedding.

Kumari soon arrived from her in-laws' house, carrying her ten-month-old son. She looked thinner, as everyone says new daughters-in-law do, but strikingly beautiful in a royal-blue shirt made from the cloth bought by Devi-maya in Biratnagar. Kumari left her son on the porch with her younger sister Myam. She then ducked through the door to see if there was any meat to eat and to mix some *jad*. Ignoring the arrival of her younger sister, Devimaya continued to sweep and to scrub pots. Every so often, she yelled impatiently to her younger siblings—Cema, Myam, and Rendha—telling them to hurry and put on their school clothes. "School is starting right *now*," she shouted. "It is time for you to go!"

A few minutes later, I was surprised to see Dhanmaya walk back onto the porch, carrying a woven bag that bulged at the sides. She removed a large leaf filled with cooked rice and put it on the mud floor. "I didn't think you would come to the wedding," she said, "so I brought some of the wedding to you."

Dhanmaya took out another leaf filled with meat and told Kumari to bring out plates. She filled each plate with rice and carefully distributed the meat, giving an extra piece or two to Kumari because, she said, Kumari was a baby's mother and needed "strong" food. Kumari immediately started to eat.

Both Devimaya and I protested that we had just eaten and did not need more food. Dhanmaya poured gravy from a wooden container onto our plates, ignoring our protests. We dutifully began to eat.

Dhanmaya sat with us for a few moments and then headed back to the wedding. Kumari climbed the soot-covered ladder to rummage through the clothes piled in a room on the porch for a newer lungi and a brighter sash. Devimaya and I followed her. Devimaya began to change as well.

"I have to go," she said. "My mother said so."

◊

It was overcast and cold. As Kumari, Devimaya, and I walked across the rice fields at the bride's house, I was acutely aware of everyone's eyes and whispered comments as we passed, suddenly realizing why Devimaya had been so reluctant to come. I had not been in Hedangna for more than two years, and when I greeted people I knew, they looked closely at my face, shook their heads, said how rosy my cheeks had been before and how thin I was, and asked if I still had no child. I wrapped my woolen shawl more tightly around my head and followed Devimaya and Kumari across the field.

Dhanmaya, who was sitting with a group of women who were mixing *jad*, saw us walk up and immediately had someone bring each of us a *tongba*. Then she found someone to bring us leaves for plates and told the men to serve us rice and meat, our third meal of the morning. Kumari started to eat. Her in-laws were much poorer than her parents, and food was scarce in her new home. She knew to eat when she had the chance. Devimaya and I picked at the rice, for appearance's sake, until someone shouted that the groom was coming. We then folded the leaves over the uneaten food, tossed them over the hill, and joined the crowd to watch the wedding procession weave through the village.

Sometime later, we sat with a group of women and watched the drummers in the courtyard of the bride's house. The groom stood off to the side under a black umbrella. The bride was inside. I did not pay much attention to what they were doing. Instead, I watched Devimaya, wondering what it must be like for her to come to a wedding so soon after the death of her fiancé. She knew that the guests were looking at her and talking about Dev Kumar, how old she was, whether she would find another man to marry, and what would happen if she did not. I wondered what she saw as she looked at the groom or how she felt imagining the bride sitting inside her home.

But Devimaya was watching the shaman Kelekpa, who was drunk and dancing foolishly in the courtyard. She had been drawn into his performance, laughing with a bit of the animation I remembered. And then suddenly her expression changed, as if she were waking from a dream. She fell silent, turned away from the crowd and the celebration, and stared across the fields and toward the Arun, as if looking into a past—or a future—that no one else could see.

◊

The day after the wedding, Devimaya and I spent the morning hauling firewood in bamboo baskets up the hill to a pile closer to her parents' house. After three or four trips, we dropped our baskets on the ground for a break. Devimaya mixed fermented rice with water from the stream, and we sat down for a snack of *jad*. We drank the beer silently for a few moments, and then Devimaya said that she had seen Dev Kumar again in her dreams the previous night. He was ahead of her somewhere on a trail and had called for her to join him. She protested, telling him that he had fallen from a cliff and died. He replied that he had not died—see, here he was, still alive. "Now come on!" he insisted. "Come quickly! Put on your clothes and join me!" So she wrapped a lungi around her hips and put on a shirt and rushed out to meet him. But then a landslide swept down the ridge and wiped out the path. She could no longer see him ahead of her on the trail. She searched and searched, but she could not find him anywhere. He was gone.

And then she awoke.

In the hot sun, after drinking some *jad,* Devimaya began to talk about Dev Kumar's death. I asked questions only to let her know that I was listening and to encourage her, if she wanted, to share more. She had been home alone on the day she heard the news. Everyone else in her family had gone to the day-long bazaar at the edge of the Arun River, but Devimaya had decided to stay at home to study for her intermediate-level exam. Early in the afternoon, a girl in the house up the hill shouted down that someone, someone named Dev Kumar, had fallen and died on his way back from the Barun *mela.* Devimaya did not know that Dev Kumar had returned to Uwa from Kathmandu and had gone to the Barun festival. She did not believe the girl and was sure that some-one somewhere had misheard the name. So she put it out of her mind and went back to her work. Sometime later, Kelekpa, the shaman and a relative, came by. He repeated the news, confirming that it was, in fact, Dev Kumar who had died. And then he stayed with Devimaya for the rest of the day. He talked about the time he had lived near Darjeeling, picking tea and working on the roads; talked about the approaching rice season; talked about Dev Kumar and what a good person he had been; talked about anything he could think of to fill the emptiness created by the news he had brought.

For the next fifteen days, Devimaya went to the forest by herself to split wood. It was good work, she said. After splitting one log, there was another and then another. At the end of each day, she was exhausted, and only then did she climb the long hill back to the village. She hated going home; she hated especially for night to come. She did not want to eat, did not want to sleep, did not want to do anything other than work. She just wanted it to be daylight so she could be alone in the woods, splitting wood.

Devimaya looked at me and paused. She was a different person from the one I had known when I was in Hedangna before, she said. Her spirit was broken now: "It would have been easier if he had died at the beginning, six, seven years ago, before I had known him for so long, before we had been in love for so long. It'd be different if I had studied more or if I had a paying job. What is left for me now?" She looked over the rough, rocky land; shook her head; and said, with a kind of hopelessness that seemed new, "The only thing left is more work."

She paused again and then began to talk about the times Dev Kumar had asked her to marry him. She had put him off because her father had told her, one night when he was drunk, that she was not to get married. She was the oldest child and the best educated. He would die soon, her mother was *lato* (dumb), and her brother was too young to defend the family's property.

Devimaya had to keep the neighbors from stealing their land. "I shouldn't have listened to him, I should have married Dev Kumar when he first asked," she said quietly. "Then, at least, we would have had two years."

We sat silently for a while, picking out the remaining chunks of fermented rice from the *jad* and looking over the terraced fields. Devimaya talked about how she wanted to die soon, so she would not have to wait so long to see Dev Kumar. "It'd be easier if I knew what it was like on the other side," she mused. "If I knew whether it was better or worse, then I would know whether staying alive was worth the effort."

She turned to me and said that if she died first, she would visit me in my dreams and tell me what it was like so that I would know. She made me promise to do the same. Devimaya then sighed and looked across the fields. She stood up, lifted the empty basket onto her forehead, and headed back down the hill for another load of wood.

◊

As the eldest child of a wealthy family, Devimaya had received at least some of the privileges usually reserved for sons. She was able to study through the tenth grade. She was the ritual sister of the one foreign woman who had lived in the area, a relationship that brought money and Western clothes into a community where both were scarce. And, most of all, she had fallen in love with Dev Kumar—the best-educated, most-clever man in the area—fallen in love in a place where people rarely have the luxury to fall in love.

Devimaya had not spoken with Kumari or Dhanmaya about Dev Kumar's death. Her friends from school were now married and had children. She never saw them anymore, and even if she did, they would not understand. She had talked to a cousin, who had not married either and was working as a scout for the Makalu-Barun Conservation Project. But the cousin had told others what Devimaya had said, so now she just kept quiet.

From the outside, I had looked with envy at Yamphu culture, wishing that I lived in a community bound by a body of beliefs and practices to support and guide me through my life. I had focused on the strength and security the culture provided. Now I wondered about the other side of that closeness, about how hard it must be to have everyone watching you, commenting on whether you were dwelling on a loss or had forgotten the loss too quickly. I wondered how it felt to have this sorrow trapped inside, reliving the memories again and again with no one to share them, no way to remember Dev Kumar as a way of keeping him alive and slowly letting him go.

While reading through my notes about Hedangna in Cambridge, I had been disconcerted to discover that, despite the details I had gathered, I seemed to have missed some deeper meaning, an invisible quality that was essential to understanding what the villagers' lives were like. But now, I realized that the time spent gathering those details was what helped me begin to understand what her loss meant to Devimaya as an individual and as a woman in a Yamphu village. I knew enough to respect and accept her silence, an acceptance that, it now seemed, in turn allowed her to speak.

Devimaya shared more as we went about the daily tasks her mother gave us to do: washing wool, beating rice, carrying loads of firewood, clearing land for planting millet. She talked about one time when Dev Kumar had passed through Hedangna. She was resting on a stone platform on her way to fetch water. He called down to her, asking her to "wander" with him, which is how young men and women courted in the village. "What would my father say?" she asked. And he replied, "Are you going to stay with your father or with me?"

When they had been in Biratnagar, Devimaya told him that since they were getting married, they had to speak frankly, and that she had heard that his mother had told him to marry a woman from Kathmandu. Dev Kumar replied that it was true, but he had told his mother that he was going to marry a woman from Hedangna, that women from Hedangna were able to work and could get by without a lot of *sukha* (comfort). When Devimaya still seemed worried, he asked her, "Are you going to marry my mother, or are you going to marry me?"

In her letters to Dev Kumar, Devimaya would write that she was going to fail the exam she was studying for, and he would become annoyed and write back impatiently, "Is this all you think about? Talk about something else!" Another time, he told her that *laj* (shame) was an ornament, like jewelry, that women wore. Once he wrote a letter in which he pretended to be sick, and she answered that he had to go to the doctor and take care of himself. Dev Kumar replied that he had seen a doctor, who told him that the medicine he needed was Devimaya.

It would all be so much easier, she said again, if he had died at the beginning, before they had fallen in love.

Devimaya and Dev Kumar had planned to walk down to Khandbari together when she went to take her exam the previous autumn. Then she received a letter from her teacher, telling her to come quickly, so she had to leave on a Wednesday, before she could get word to him in Uwa, forty minutes to the north. Dev Kumar came to Hedangna to meet her on Thursday, as planned, only to find that he had missed her. They did not see each other for several months. And even though both of them had been in Biratnagar for an entire week, they saw each other for only two afternoons.

I think of all the times they missed being together, because of Devimaya's shyness or deference to her parents, or simply because of the circumstances of being a young man and young woman in love in a place with little privacy and no way of communicating other than handwritten letters. I think of the times they did have together, moments from the past that are now places in the present, to which Devimaya can return to remember what she has lost.

◊

The afternoon before I left Hedangna for Khandbari, I joined Baiseti Thuma in her millet field just above Dhanmaya and Dilli's house. As I helped her gather rocks into a small pile, we talked quietly about her life now that she had moved in with her daughter. She asked about my work in America and about Brian.

After a while, Baiseti sat back on the hard stalks of cut millet and asked about Devimaya. She said that she worried about her and wondered if she was doing all right. It will take time, she said, and paused, looking down over the village, brown and drab on an overcast day. And then she began to talk about her own sorrow when her eldest daughter had died, suddenly, in childbirth when she was only twenty-one. For months, a year, afterward, the grandmother said, her heart had been dark with sorrow. And when the heart is dark, she said, the eyes cannot see. She would head down the trail from her house, only to reach the bottom and remember that she had intended to go up the trail. Or she would climb the hill, only to remember that she had meant to go down. When people fed her, she would eat; otherwise, she forgot that she was hungry. It was only slowly, after a year, after two years, that her heart again became light. And only when her heart was light, she said, were her eyes able to see.

◊

Dev Kumar often came to Devimaya at night in her dreams, where his presence was more palpable than it ever had been when he was alive. But then he vanished the instant she awoke, and she was left lying awake for hours, in the dark, with nothing but his memory and the weight of his absence, nothing but the weight of a future stretching forward without him. For the first month, she thought of him constantly, she said; then after a few months, she found that she became absorbed in her work and that hours passed during which she did not think about his death or her sorrow. With relief, she believed that the edge of grief had finally passed. Then a gesture or smell suddenly brought him back to the present. Sometimes, these encounters left her more acutely alone, more aware of her loss, overcome, again, with longing for what had been and could have been. Other times, she simply was angry with him for having left her, to live her life on her own. And still others, she was overwhelmed with the depth of her love.

Shamans and priests who can see what is invisible and what is visible move at will between worlds that the rest of us can travel through only in time—in the time it takes, as Baiseti Thuma said, for a heart to again become light. The healers do not pretend that their ability enables them to keep the dead out of their houses and out of their dreams; they never suggest that, because they can communicate with the nonliving, they are not haunted by memories of the past. But in a ritual, the healers create a clearing outside linear time where the boundary between worlds is less sharply drawn, where the feeling of an invisible presence, the sensation that Devimaya had again and again of Dev Kumar's hand on her shoulder, become as real and substantial as objects we can literally touch. In the opening created by the ritual, we can see and feel, if only for an instant, that what has been lost to the physical world has not been—and never will be—lost to the world.

And yet, even if we know that the sudden feeling of the invisible presence of someone we have loved and lost is as real as the visible, material world in which we live, how can we hold onto that knowledge once the ritual comes to an end? Once the experience passes, once the clearing disappears, how can we not feel, again and again, the longing to have what has been lost to this world, here, now, in this world?

I always wondered—but never asked—what it is like for the priests and shamans after they return from their journeys, once a ritual is complete. Are they relieved to have found their way back, and do they cross through the *tsawa* quickly, without a moment's thought about the world they are leaving, eager to return to the realm of the living? Or do they hesitate, lingering in the shadows and looking through the *tsawa* at the flames flickering across the lined, worn faces of the men and women gathered for the ceremony, looking as one does at a brightly lit stage, a world somewhat less real than the world where they stand? Once they return to this world, once they can again hear the voices and touch the skin of those they left behind, do the priests and shamans forget where they have been? Or do they also feel a sense of sorrow, not for what has been lost—because they know, from the perspective of seeing "double," that loss is never absolute—but for having known the "capacity for 'living easily' which is the hallmark of those few beings who know that they will live forever" (Calasso 1993:96) and for having known, equally, that that capacity is not and never will be for them?

◊

Devimaya and I traveled to Khandbari together three weeks after I arrived in Hedangna. I was heading home, and she was going for an interview to become a teacher. Dhanmaya arose long before dawn to prepare food for our journey. The previous night, she had told us that she would accompany us as far as Gadi, where she had to sell rice to pay off a loan taken to give Devimaya money for her trip. But as she spooned rice and greens onto our plates by candlelight, she said that she could not come. We would continue to head south, she said, while she would have to return alone to an empty house. It was easier to stay behind.

We left when it was light enough to see the trail, walking quickly and not stopping until we reached a stone platform well above Num, several hours later, where we ate the curried potatoes and drank the millet beer that Dhanmaya had packed. We set off again, and by the time we arrived at the ridge above Khandbari, ten hours later with two hours still to go, both of us were exhausted. We had not talked much, but now, as we walked down the hill, Devimaya asked me to say something to keep her mind off her feet, which hurt. I said that I hoped that she would find some comfort in Khandbari, that she would not be too homesick, and that she would not miss Dev Kumar too much. "Before, all I could think of was Dev Kumar," she said. "All I could do was miss him. But now, you are leaving. Now I'll miss you as much as I miss him."

The following evening, the night before I left for Kathmandu, Devimaya and I went to a tea shop for a meal of *tongba* and goat meat. The electricity was off, so we sat close together at a table in the dark, with our *tongbas* and meat, as though we were the only patrons in the shop. Devimaya wore a baggy, white T-shirt over her lungi, making her skin and hair seem darker than usual. She looked beautiful.

She placed her hands on the table and dropped her head between her arms and kept it there. I asked what was wrong. She replied that I was leaving the next day and that she was sad. "But I haven't left yet," I said. "We have to talk now, while there is still time; have to say things so that we can remember them later, when we are no longer together." Devimaya sighed and looked up. She ate a piece of goat meat and took a sip of *tongba*. She talked about when she first met me and how shy she had felt. She still felt shy because she cared what I thought of her. But she spoke up anyway, she said, even though she was embarrassed. We then talked about other things: Dhanmaya and how generous she was, and Kumari and how she had changed now that she was a daughter-in-law. We finished our *tongbas* and then went back to Raj Kumar's house, where we stayed up late drinking *raksi* and telling stories with his wife.

The next morning, I had to walk to Tumlingtar to catch the plane to Kathmandu. After drinking a cup of tea, I lifted on my pack and said good-bye to Raj Kumar and his wife. Devimaya, who had come from Amrit's tea shop, where she was staying, said that she would walk with me to the edge of the bazaar. We walked silently, Devimaya staring at her feet and me wondering what I could say that might make her feel better. I began to talk, not because I had anything left to say but because I wanted to fill the space of the silence, wanted to give her words to keep her company, later, in her loneliness.

I told Devimaya how lucky we were to have had this time together. I said that some people, like Kumari, seemed content to stay in a place like Hedangna. She belonged there, but both of us were looking for something else. We often would feel lonely and sad, while people like Kumari and Dhanmaya might never experience such loneliness or understand what it is like. I told her how much I cared for her. Devimaya kept walking, eyes on the ground before her feet. I asked if my words made any sense. She said that she understood, but that she could not say anything now. She was just too sad.

Devimaya stopped abruptly when the path forked and told me that she had to go. She brushed her hand across her eyes, as she had the night I returned to Hedangna, and, without saying good-bye, turned and headed back along the path. I ran after her and grabbed her rough, stubby hands in mine. I held them tightly and pressed a piece of polished green glass in the palm of one hand. It felt so pitiful and small, this bit of colored glass, but it was all I had. I told her that I had been carrying this glass with me and that whenever I felt sad or lonely, I rubbed my fingers across its surface. I told her to carry it with her and to do the same, whenever she felt that way. She nodded and said, again, that she just could not talk. She turned and headed up the trail, without looking back.

◊

Once while in Hedangna, I dreamed of a close friend with whom I had lost touch. We had not parted on good terms. When I awoke, I realized how sorry I was that I had not been in touch, how sad I felt to have let a friendship that was so important slip away. The next day, I walked for half an hour to a shop in

Gadi to buy some kerosene. When I passed the small hut that was the post office, I greeted the thin, old man who was sitting behind a small wooden table
and sorting the mail and asked, as I did each week or so when I came to Gadi,
whether I had received any mail. His answer was always no. But that morning,
he said, "Wait, yes, you do have a letter!" With surprise, I saw that the letter
was from the friend whom I had seen in my dream the previous night.

Later that day, I told a woman in the village about this incident, noting how
remarkable I thought it was—the strange synchronicity of things. She did not
seem surprised, but said, simply, "Of course. Your minds must have met in
your dreams."

◊

I tell myself that even though Devimaya and I do not write to each other, even
though we have not seen each other since we parted in Khandbari, the times
I think of her are just as valuable, just as real, as a letter might be. But dreams
are conscious messengers for the villagers: they are just as real a way of communicating—or, rather, just as unreal—as a letter. I come from a world that
has severed its connections with these other ways of knowing, from a culture
that considers such signs, when they are even noticed, as suspect, dismissed
as nothing but an interesting coincidence or an overly active imagination.

We make pilgrimages to sacred places, to places where gods have made
their mark on the land. These places have a power and a presence that cannot be seen with the naked eye. But the sacred not only is present in far-away
places, but is a quality we experience when we open to the world around us,
to the sacred spring that flows through all of our lives, if only we know to perceive it. In my friendship with Devimaya, I came to see that my ability to experience the sacred depended as much on my relationship with a person or
place as on the person or place itself. With Devimaya, I experienced a quality
of openness and connection that I had encountered only on pilgrimages or in
village rituals. It was a connection marked by loss; perhaps that was what allowed us to experience this quality of the sacred. We knew that there was not
time to be any other way.

19
MANGUHANG

March 1997
KATHMANDU

I had just ordered a grilled cheese and tomato sandwich and a salad; then I would eat fruit and yogurt; in an hour, I was going to have sag paneer and chicken tikka and a nan or two for dinner. I was wearing clean clothes and earrings, and was very, very tired. I had not landed yet in Kathmandu, was still with Devimaya in Khandbari, wondering about her day and where she would stay the night and whether she, too, felt lonesome and lost, still worn out from the long walk the day before. I thought of the idleness of her days in Khandbari, of how hard that idleness always was for me, waiting for time to pass, waiting for some distraction to fill the space of the waiting. I wondered what she was waiting for.

The next day. I was sitting alone with a beer at a small table in a restaurant in Thamel, the tourist section of Kathmandu, waiting for my food. I was invisible, watching people who could not see me, on the outside looking in—the heightened perception that comes from having been so absorbed in events in

a far-off place. I thought about love and loss and the traps we get caught in because of the mountains that block our vision, and about what happens when
people go beyond that ridge of mountains, crossing the boundaries of what
they know, entering a world about which they have no stories, why they cross
the borders, and what they do once they have.

Robert had left me a letter, saying that he was in town and would be in
touch. That night, I tossed and turned in the tiny bed in my stuffy room, wondering why he had not called. Tempos rumbled by on the street outside my
window. Crows squawked. I began to hear the sounds of early morning in
Kathmandu: water pouring from windows, horns honking, Hindi songs blar-
ing. I gave up on sleep, got out of bed, mixed a cup of Nescafé with powdered
milk, and sat down at the tiny table to work. When it was finally late enough
for the restaurant to open, I went down and ate white toast with bright-orange
Indian-made marmalade. I felt like the man in the film *The Man Who Planted
Trees,* in which whole images were rendered with a series of hand-drawn lines:
a jawbone and nose for the face of a man. The effect seems to capture all that
cannot be captured in a sketch, just by its incompleteness. Part of me, too, felt
as though it were fading into the space beyond, my aloneness the only thing
that seemed real. At home, the props—people and tasks and places—hold me
in this world. Without the props, the only thing keeping me here was a sheer
act of will.

I went out on the street to pass the hours stretching before me, trying to
create a distraction, filling the silence of the telephone, leaving the room because I could not bear to stay inside. Kathmandu seemed seedier and grungier
than before. I was more aware of the crumbling old and cheaply constructed
new buildings. The Maoist rebels were killing people in the west; the chief of
police of the country recently had been murdered, and his death was not mentioned in the newspapers. The naive hope—held not just by me—was that
life could go on, that Nepalis would stay happy and content even as they increasingly felt their poverty, even as the promises of politicians, development
workers, tourists, and teachers kept being broken. The things that brought
Westerners to Nepal—stunning mountains, peaceful monks, friendly villagers—were very different from the things that absorbed a nation struggling to
get on its feet. Exhausted from the crowds and the sooty black air, I returned
to my room and tried, again, to translate Kelekpa's *mindhum* describing his
descent into the underworld. I kept staring out the window.

In Hedangna, the villagers name the spirits from the past—Matlung
Thuba, Chaketangma—and invite them into their homes, where they feed

them *jad* and new rice. In Marilynne Robinson's novel *Housekeeping,* Sylvie let the wind and rain blow through the house, allowed the grass to grow, and wandered through the orchard at night, trusting that the cold had as much to offer as the warmth, if she let it. She knew that the only way to free herself from the shattering of images inside was to turn into those memories and let herself be consumed. That turning in could destroy a person's home or, as with Sylvie, force her from her home, but that was better than being haunted.

Finally, there was a knock at the door. Robert and I had not seen each other and had rarely corresponded for two years. Although he was the one person I wanted to see, I was determined to keep the wall between us intact. And yet once again I found that what I wanted more than anything—the sharp recognition that this moment is all that endures—what I have never found with another person, I found here, in this moment, doing what I swore I would not do.

As he left, he promised to call. I shut the door, turning back to my desk. I sat down and tried to work. Five minutes later I turned, reach for my bag, and headed back to the streets.

◊

It was 3:00 A.M. Tossing and turning, longing for daylight. Was that all it was? No longer taking it for granted that morning would come. Shaking my bamboo wand again and again until a light appeared on the horizon.

He did call. And, later, he returned. What made for this intensity of presence? When I am not here, it is as though I never will be again—and then I am, here and nowhere else, in this place where the world drops away and all that exists is this moment and the push to make it more present, to make the present as palpable as I can. It is a this-ness that words cannot touch: not something to own, not something that can last. Does it then become too much? Must I drift away, back to solid lines and firm ground? Or can I stay, forever, in the "quick now, here now, always" (Eliot 1971:59)?

We arose in the early morning and walked through the dark, just-waking streets of Thamel, stars filling the sky, to his motorcycle and a future that must be kept free from any expectations.

◊

The days passed, and the ground slowly began to hold me up. Nepal seemed more like it had always seemed. The carpets and tables and cups and saucers offered a sense of comfort, even as I still felt a sense of everything crumbling. I wrote a brief report about grazing rights under the *kipat* system for the Makalu-Barun Conservation Project. I translated Kelekpa's *mindhum* and found a Nepali to transcribe the tapes I had not had time to finish with Raj Kumar. I spent time with Tibetans whom I knew from my work in Nepal in the past. I became absorbed in these tasks even as they were set against a bigger backdrop of waiting, waiting for the telephone to ring, waiting for a knock on the door.

Robert stopped by one or two more times, never for very long. And then two weeks later, I returned home—to Brian, to another alone space, alone with the weight of what I had done, the uncertainty of what was to come, and the distance with Brian, a distance that I had created.

Another anthropologist told me that, just returning from Nepal, I was "a delicate emotional structure" and that I should linger, move slowly through this place because it was a time when all categories were suspended, when I could see things more clearly, without the blinders of my culture. In these in-between times—no longer of the place we have left, not yet of the place we have reached—we can understand things that we cannot at any other time. He told me to treasure it, but I just wanted time to pass, wanted to be back in the past or safely into a future where I had forgotten all I now longed for.

◊

April 1997
CAMBRIDGE, MASSACHUSETTS

Back in Cambridge. I met Julie and Denise, her sister, at the oncologist's office to keep Julie company during another round of chemotherapy. Julie was a friend from Harvard. Both of us had lived overseas for several years before entering graduate school; we shared a desire to know things deeply and fully, and we quickly had become close friends. But we were also anxious in similar ways, anxious about our work, about school, and about our futures—an anxiety of our own that was exacerbated by the environment of a doctoral program at Harvard, where what mattered were our ideas, our performance, and the future. We expressed our apprehension similarly as well, both of us becoming tight and controlling, tense and competitive. Neither of us had many skills to manage our nervousness. Julie's anxiety made me anxious, and so I had begun to avoid her, making excuses without acknowledging what was wrong.

And then we had left to do our fieldwork, me to Nepal and Julie to Chile, where she worked in a fruit factory, exploring the impact of globalization on fruit production.

While in Chile, Julie had found a lump in one of her breasts. At a local hospital, she was told that it was not cancerous and could easily be removed without returning to the United States. Julie was a hard, disciplined worker, not one to give up easily. She had the surgery and then went back to her work in the fruit factory. A year later she returned home, only to discover that she had breast cancer, which may or may not have been diagnosed had she come back earlier. She underwent more surgery and began a round of chemotherapy and radiation.

Both of us had returned to Cambridge humbler, quieter in the face of things we realized we could not control. We began to explore ways of being present in our own lives and with each other, other than what Harvard had taught us. We became friends again. I took her food shopping, cooked her meals, and drove her to the hospital for test after test. And then the cancer went into remission. Our conversations shifted from mortality and chemotherapy back to fruit pickers and farmers. But our friendship was different. We were more forgiving of each other and of ourselves. We had a clearer sense of what mattered to us and cared more about being a good friend than a good student. And then a year after her remission, Julie discovered that the cancer had metastasized to her liver.

The small room where the medications were administered was crowded with men and women, each sitting in a dentist-like chair, an IV at his or her side. A plastic bag attached to a pole was filled with a clear, innocuous-looking fluid, which dripped into a tube that ran through a needle and into each patient's veins. Everyone in the room seemed old and frail; any hair that they had was gray. The nurse gave us a separate room so we would be more comfortable and, I think, because we were so much younger than everyone else. After Julie got hooked up to an IV with her own plastic bag filled with fluid, we took turns reading from one of Julie's favorite books, *The Little Prince*. Julie read the part about the little prince bidding farewell to the fox he had tamed:

> "Goodbye," said the fox. "And now here is my secret, a very simple secret: It is only with the heart that one can see rightly; what is essential is invisible to the eye."

"What is essential is invisible to the eye," the little prince repeated, so that he would be sure to remember.

"It is the time you have wasted for your rose that makes your rose so important." (Saint-Exupéry 1971:70)

I read the part about the well that the pilot and the little prince found in the middle of the Sahara:

"I am thirsty for this water," said the little prince. . . . I raised the bucket to his lips. He drank, his eyes closed. This water was indeed a different thing from ordinary nourishment. Its sweetness was born of the walk under the stars, the song of the pulley, the effort of my arms. It was good for the heart, like a present. . . .

"The men where you live," said the little prince, "raise five thousand roses in the same garden—and they do not find in it what they are looking for."

"They do not find it," I replied.

"And yet what they are looking for could be found in one single rose, or in a little water." . . .

And the little prince added:

"But the eyes are blind. One must look with the heart." (78)

Julie moved her chair, so she would not have to turn her head to see Denise and me at the same time. We finished reading, and she told me that on the way to the doctor's office, she had told Denise that either she was driving too quickly or Julie was moving too slowly because they were arriving at places before Julie could find them on a map. It was easier to joke about the effects of the cancer and the chemotherapy than to talk about what was really going on or even about what we had just read. And so I said that I always tried to walk more slowly when I was with Julie but that once, when we were really late, I had picked up the pace just a bit and Julie had said, "Oh, so now we're running!"

Then we talked about having babies and her sister's future and shopping at the mall. We laughed and laughed until tears came to our eyes, for no reason at all, and the nurses and doctors kept coming into the room to check on Julie and to see what was so funny.

Julie's cheeks went from bright-red spots to greenish-yellow smudges in the two hours we were there. The catheter, which had been inserted below her collarbone to administer the medications because the veins in her arms had been used too often, protruded more than usual because Julie was so thin. A nurse kept entering the room to check it, each time saying too loudly that we should not worry, that it would be okay.

◊

Julie once talked about how glad she was that she had not spent the past year worried that the cancer would come back; at least she had had that time cancer-free, to live then and now, to remember. I thought about this awareness of time and how it can cast a shadow over the present, or not.

A few days after her chemotherapy, Julie called to say that her oncologist had told her that the medications were not working because it was difficult to attack both cancerous sites. I listened numbly as she spoke, not knowing what to say. I realized how much I had been counting on a different future, how much I had imagined she had a future. And I realized how much the belief in a string of moments following this moment affected my presence—and my absence—now, in the one moment I could really count on, cancer diagnosis or not. The diagnosis, like my transgression with Robert, simply heightened that awareness.

◊

Brian and I went to the beach north of Boston and lay on the sand, absorbing the warmth of the early-spring sun. I felt myself slipping away and thought that I should talk about it, let him in, share where I was going. Yet I could not trust myself to express the depth of my aloneness and could not trust him to listen, to try to understand when my words failed. The distance between us was too great, not just because of my time in Nepal and everything that had happened there, but also because of Julie's illness. Julie and Brian were not very close. They were too different to have much to share. Brian focused on doing things in the world, on making a difference that was visible. He did not understand what Julie had done and thought that she, like me, was lost in the process of searching, unable to produce anything, particularly anything that, in his estimation, mattered.

But the distance between Brian and me was deeper than that. I think that Brian believed that if he could figure things out, if he could just get everything right, then everything would be all right. The one thing I had learned in Nepal was that you could not figure things out, that you never arrived at a place where everything made sense. Everyone just did the best they could.

I knew that this was true, one of those truths beneath all other truths. And I knew that it was a difference between Brian and me that, in some way, we had to reconcile or make peace with if our marriage was to last. Yet I could not risk ending up feeling more lonesome than I already felt. And so I pulled back. We dozed off. After waking, we walked along the beach in the late-afternoon light and then headed back to our car.

Now, more than ever, our relationship was marked by silence. Not the silence when there is no need to speak, a silence that is comforting and nourishing. But the silence of something hidden, a silence that is more palpable, more present than the words we expressed.

◊

I had to get on with my work, with turning my dissertation into a book, which I had received a grant to complete. I began to read about mystic journeys and sacred places, about how it is only an illusion that we lose what we love, that if we just reconfigure how we think about space and time, everything is present all the time. Now is never then. Now is always now. Even then is now. The shamans in Hedangna ultimately do not go anywhere on their journeys, or, rather, their destination is *here,* this moment. They are in the moment as fully as they can be, seeing what they can control—their presence of mind, body, and spirit—and what they cannot: whatever comes next. And it is this absolute presence that lifts the veil and lets them enter the world of the ancestors.

I thought that I could come into my work smoothly, writing about the unseen world through which the shamans travel. But I could not. No matter that Devimaya can see Dev Kumar in her dreams. No matter that in the other realm, we all "see double" and there are no distinctions between the living and the dead, between absence and presence. In this world, there are.

And it is in this world where we have to live, most of the time. And so Devimaya brushes her hand, again, across her eyes and carries on, but with a loneliness inside that she had never known, a loneliness that will become less acute but will never disappear.

The shamans in Hedangna do not try to help people transcend the material world. Instead, they help them deal with the consequences of their immersion in this world, seeing that grief is just the other side of love and understanding that we always lose what we love and that all we can do is continue to live and love until it is our turn to be the one who is lost. The shamans help heal the suffering caused by immersion in this world by traveling deep into it—to Manguhang, by the lake at the center of the earth—not by rising above it. And in this realm of the ancestors, they receive a flash of insight that is healing.

◊

Through e-mail, I was in touch with Robert more than ever before. Our messages closed the physical distance between us by creating a quality of attention that could not have been sustained had each of us actually been present in the other's world. The focus burned a tunnel through space and time, creating a feeling of dislocation similar to that of being alone in Kathmandu, similar—I wonder—to that of journeying to Manguhang as the firelight flickering across the faces fades; all I saw was the longing leading me along, deeper and deeper into the unknown. I always held back before. And yet here and now, with Robert, I was invited to go farther and farther, deeper yet, handing myself over to the longing to dissolve all distance between us.

And then I found myself in a clearing where that distance—half a world—suddenly, was gone. I felt a presence beside me so strongly that I was afraid someone would notice. A feeling of awe: walking, walking, watching, watching. I was not alone—even as, physically, I was completely alone.

In *The Divine Comedy,* Dante says to Beatrice, now you know the power of my love for you, treating shades as if they were solid. It is not good to talk about, touching an ache that is so deep, beyond sadness, an ache that precedes every other feeling. I suddenly was no longer afraid that Robert would not return, no longer concerned about the lack of a future; even the weight of sadness and loss had lifted. My fear was that I never would arrive at this place again, that every other experience would be marked by the desolation that it was not this: absolute presence, an end to limit, an end to time. I saw that fear, even as I knew it was my fearlessness that had brought me to this clearing, which I now never wanted to leave.

They journey into the world beyond the veil of what our eyes can see. Then they return, the priests and shamans and poets, and give us words and art, scratches in the sand that lead others down, drawing us along. I always knew that I was being led somewhere, but never knew it in this way, never knew that this was where I would find myself. I felt as though I finally had arrived at a place I had been searching for my entire life without knowing what it was.

By transcending the fear of losing what we love, by not running from the roaring lions at the gates that Kelekpa passed through in his descent into the underworld, we arrive at a realm that is beyond death, the place of inextinguishable laughter of the gods, the clearing in the woods where what we encountered endures.

When I went to Nepal, when I spoke with the shamans and priests in Hedangna, when I opened the door each time to Robert, I was searching for something that I could not articulate, even to myself. With Robert, I unknowingly stumbled into the clearing that I had studied as an outsider. In that moment, I knew without question that I, too, could "see double," knew that there really was more to the world than we can see and touch, than we can know with our minds. And now, having experienced that erasure of time and space, all I wanted was more, wanted again and again to see the visible and the invisible because that experience felt more permanent and more real than anything I had known.

My journey to this place was not as smooth or as straightforward as I had imagined that of the shaman or of Minaba and Sepa to be from a distance. Had I known how difficult it would be, I doubt that I would have left home. But we cannot know at the beginning of our journey what we will encounter along the way; we can only follow the longing inside—the longing to discover the source, to find the essence in grain, as the Yamphu say, that makes it last; the essence in life that makes a life worth living.

Once I had arrived in that clearing, if only for a moment, it seemed as though I had two choices: either dismiss my experience as a sign of an overly active imagination or accept a world in which such encounters were possible. I knew enough about being in a world that dismissed the realm of the sacred. I wanted, instead, to discover where I might be led were I to believe that what had happened was true.

◊

Each day, I ran through the streets of Cambridge to a stone bench in the middle of a planted pine forest next to Fresh Pond. The bench is said to have suddenly appeared one morning, an anonymous gift in the night. I came here to run, but I also came this way, to these woods and this pond, because of the stone bench, on which Virginia Woolf's words from *Orlando* were carved. Each time I came, I paused long enough to trace the final letters with my finger:

> My forehead will be cool always. These are wild birds' feathers—the owl's, the nightjar's. I shall dream wild dreams. . . . I should lie at peace here with only the

sky above. . . . Indeed, she was falling asleep with the wet feathers on her face and her ear pressed to the ground when she heard, deep within, some hammer on an anvil, or was it a heart beating? Tick-tock, tick-tock so it hammered, so it beat, the anvil, or the heart in the middle of the earth. (1995:122)

I then lay on my back on the bench, looking up at green needles against blue sky crisscrossed with black crows, and imagined the heartbeat at the center of the world, imagined that it was the beat of my own heart, deep inside my chest, never stopping the dance, whether I knew it or not.

◊

Julie's next chemotherapy session was different from the previous one. Her mother and stepfather accompanied her this time. As we settled into the room, Julie realized that she had forgotten her bag and asked her stepfather to get it for her. He was gone for fifteen minutes, and when he finally returned, he brought the wrong bag. He said that he would go again, but Julie assured him that she did not need the bag. Really, it was okay, she said. And then she assured her mother, who was crying, that everything would be fine. Julie often was hesitant to be too open, too honest about what was going on, especially about how she was really feeling and especially with her mother. This made sense. This was her mother. Julie was her daughter. And Julie was dying.

I kept my copy of *The Little Prince* and Barry Lopez's *Crow and Weasel* in my bag and instead watched the chemicals slowly drain from the plastic IV bag, through the catheter, and into Julie's vein, praying that this poison would let her pave a pathway into the future.

◊

Julie came for dinner. She chopped the garlic into maddeningly tiny pieces, although what I needed was garlic cut any way, ten minutes earlier, and told me that she had decided to stop the chemotherapy and wanted to talk about her funeral. She did not want to survive this recurrence of the cancer because the cancer would metastasize again and again, and she did not want to deal with it anymore. She talked so matter-of-factly—her eyebrows beginning to grow back, the edges of her hair sticking out from under her purple, blue, and white scarf. She had bought a wig but never wore it, instead wrapping brightly covered scarves around her head as a way of acknowledging what she was going through but still feeling beautiful. At home alone and with friends, she wore a little black cap.

I said that it all seemed unreal. She agreed, because she did not feel sick right now. And this was what she wanted, she said, to feel well, not to go chasing after a cure, especially a cure that made her feel worse.

She was quiet for a while, as she continued to chop the garlic. After a few moments, she said, more quietly, that she was afraid of the pain, afraid of depending on people, afraid of becoming someone she was not. She continued to chop. And then she said, the resolve back in her voice, that her fear was the clay from which to make her pot.

◊

While studying anthropology and while first living in Hedangna, I believed that I eventually would understand enough to say something conclusive about another culture, or at least about the women in that culture or the children or the educated men, or even something conclusive about another person. I was looking for a place where I could gain a bird's-eye view of the cultural landscape. Instead, everywhere I turned, I discovered ambiguity, fields and rights overlapping, truths that looked different depending on where I was standing. This awareness was deeply unsettling, not simply because I began to doubt my reasons for having gone to Hedangna in the first place and my ability to communicate what I had discovered, but because the one truth I had been raised with was that there was a truth and that I should tell it.

When I tried to write about Hedangna—about Devimaya, Dhanmaya, Raj Kumar, or Amrit—I often thought about how, as an anthropologist, I might explain the choices I had made. Everything that seemed to matter most in my life, all the decisions I was making, emerged from the most intimate and private moments of my life, from things beneath the surface, the things I chose not to share with others, particularly outsiders. I thought of the different voices speaking to me at each step along my journey—those of Brian, my mother, my father, my sisters and brother, my advisers, and my friends; those from books I had read; those of Devimaya, Baiseti Thuma, and Chute Rai; those from my experience as a witness to Julie's journey—all these voices clamoring for a say in each choice I made. And, most important, especially in understanding my unexpected decisions—going to Nepal, climbing into the cab in Kathmandu—I thought of my own voice struggling for a say amid the rest. That listening to my voice was so difficult, yet felt so essential, that speaking not simply in private but in secrecy was easier, was connected to my personal biography, not to anything meaningful that might be said about my culture. Certainly there were things I could say about my place in the social structure, about my education and my class, or, to be slightly more specific, about where I had grown up and had and had not traveled. But none of that revealed very much, or at least not the things I was interested in knowing. All the voices, each expressed and repressed for different reasons, were what mattered. How could anything so particular to my own journey ever add up to anything meaningful that I might say about my culture?

I thought of the shaman's journey to Manguhang, a journey into the heart of the world. I wondered if I might find what I was seeking by going more deeply into life, into each moment—into the details—rather than stepping back and removing myself from the context and searching for a vantage point from which I could perceive the whole. I wondered if the essence I was seeking—an essence that transcended the particular—could be found only in the particular.

◊

A week after Julie came over for dinner, I returned to Brian's family's cabin in Canaan, New Hampshire. I was alone. I shut off the motor of my noisy truck and slammed the door, and suddenly it was quiet. I put wood in the woodstove and checked the bulbs I had planted the previous autumn, which, amazing of all amazements, were beginning to sprout. I walked through the woods, did yoga by candlelight, and burned dried flower petals in a shell. Then I cooked brown rice and beans, filled a wooden bowl, and sat cross-legged in the shadows by the woodstove and watched the steam rise from the food, drinking a cup of nettle, red clover, and mint tea.

For the first time, I understood what people meant by paying attention while washing dishes. In that moment, everything felt so peaceful and full, just that moment: the candle burned, the fork moved, the shadows gave the light its meaning. Everything else was so uncertain. Why rush? Just stay here. Light on fork. Beans and rice. The only thing I could control: whether I was here, in this place of stillness, or not.

A friend once said that there is a whole realm of spirit beings who communicate with us in different ways, whom we have cut off from our consciousness. The trick is to begin to learn to move in and on a surface that is both visible and invisible, where what we see with our eyes is not all that matters. I thought again of the wooden bowl and walking stick that led Minaba and Sepa to their new home. I wondered what it would take to trust that my own bowl and stick would appear, wondered if this time with Julie and with Robert, this time of learning to let go of a future with either, was to make sure I really knew—with my heart and not just my mind—the fear that they would not.

◊

I drove back to Cambridge. With Lida, another friend from graduate school, belly bulging with a baby two weeks overdue and contractions just beginning, and Julie, bald from another round of chemotherapy, which she had decided to continue. I went to the Brattle Theater to hear Anne Lamott. She talked about how characters choose their authors. They come to you, she said, so you can tell their stories. I thought of Devimaya, Baiseti Thuma, Chute Rai, and Raj Kumar. "Your task as writers is to look at them, to look and to look and to write what you see. Don't try to be clever," she said. "Just write what you see." And then she added that nothing anyone writes makes any difference, that there is no need for another book to be written, but that she could not think of anything she would rather do, the gift she felt of being one of the storytellers of our century.

As we stepped outside, Julie reached her arms through each of ours, and, arms entwined, we walked down Brattle Street. I remembered the cold, early-spring morning when Lida and I had gone with Julie to meet with her oncologist. Waiting in the patient room, we took turns climbing on the scale, inadvertently bringing attention to our bodies, the one thing we had tried to avoid. There was nothing we could do to change the fact that Julie had cancer and we did not. And so we offered one another the only thing we had to offer: our own presence of body, mind, and soul. We linked our arms and walked down the street, placing one foot in front of the other.

I began to see the gift of opening to these friendships that acknowledge how confusing and difficult and big everything is and that recognize that all we can do is keep one another company. I began to understand that there are other ways of treating one another, our selves and our bodies, than those we are taught by the world.

◊

In Canaan, I wrapped myself in a woolen blanket and sat on a rock in my garden in the night. A sliver of moon, in a sky studded with stars, hung over the trees. Lavender poppies and low-growing clumps of chamomile surrounded me. Branches reached across the black sky. I thought about how hard I had been trying to bring the darkness into the light, now finally beginning to understand that the point is to let darkness be what it is—healing, nurturing, soothing, and embracing, all the things I could not get from light. I began to see what emerged from the darkness once my eyes adjusted, once I began to trust that I could see and that there was something to be seen. At first, it was simply black. Then I saw shadows, although it was no less dark, and shades moving past trees. I had been trained to believe only what was visible, to think that I had only what I could hold. As I embarked on this journey of trusting that people make themselves known—live in me in other ways, not just in flesh—a pilgrimage of sorts, I began to develop my own relationship with darkness, the darkness of Kelekpa's descent underground. I realized that my journey to the Khembalung caves had prepared me by showing that by stripping away, by letting go of the things I thought I needed most, I would receive what I longed for: the quality of presence and connection I had set out to find.

I realized that I would not change where I was, would not step back to those late-afternoon hours in a room overlooking the streets of Thamel. But I would, of course. How could I not hunger for that passion and presence, even as I knew that it could not last, not because I wanted to be where I was, which was a hard, lonesome place, but because of what I now knew that could not be taken away, knowingness, a knowing that does not depend on anyone, that is in the way Dhanmaya holds her head and in the wrinkles on Baiseti Thuma's face. Part of the sadness is about growing up, realizing that life is other than what I thought it would be. But also realizing that it is okay down here on the wet ground, richer, more alive—more filled with reverence and joy than I ever imagined it could be. Bare feet on wet earth.

Weeks later, Julie called to say that she sometimes worried about how much time I spent alone in Canaan; she said that I was doing what she always did: trying to hold on to something outside myself, trying to control what could not be controlled. The meditation, she said, was to trust that when I loved as deeply and wholly as I had, that love was mine. I became that love, and the love that was me could never be lost or taken away—even when I lost what it was I loved.

◊

Willing, like Ike in "The Bear"—who sets aside his compass and gun—to trust nothing but our own presence of mind, body, and spirit, opening to what cannot be controlled, we find ourselves in a clearing. On the far side of the clearing is the bear.

And then it is gone.

PART FOUR
birth

20

BIRTH

I n "The Bear," the rest of Ike's life is an unraveling from the moment in which he encountered Old Ben in the clearing in the Mississippi wilderness. The first time I read the story, I thought that was the only option, that having encountered the sacred, Ike had no other choice but to renounce the history of exploitation—of the land and of the people from whom he descended—and his farm and family, living the rest of his life alone in a one-room apartment. But was that really true? Was it also possible to open deeply enough to encounter the bear and still find a way home?

◊

June 1997

One afternoon in the early days of summer, I climbed the hill behind the house in Canaan; passed through the back field, where the sheep grazed; and walked along a leaf-covered trail into the woods. I walked slowly, checking on what was growing, enjoying the warm air. As I walked, I gathered fallen leaves, acorns, and small stones flecked with mica. I reached a low, flat rock, like an altar, at the top of the hill; kneeled on the moist earth; and cleared away the dry leaves and flowers I had put there a week earlier. I carefully placed the new items on the rock, and then lay on my back and stared at the sun-lit leaves, at the shadows, at the green leaves against blue sky. "Sun unborn mars beauty," James Joyce writes in *Ulysses,* a line I had loved as an English major studying these ideas from a distance. We can no more stare at the sun than stay in the darkness. We cannot remain in the clearing where the shamans go. The goal, it seems, is to create that moment again and again in the present. Yet it was so different here, lying on dead leaves, feeling the sharp roughness of quartz under my back and watching the shadows glance over the green leaves above. On the outside, with no sense of how to find my way back to the clearing. But this is where I am.

◊

In June, Brian went to the Colorado River, near the Grand Canyon, to photograph Paul Winter recording a new album. I joined them for the last night, renting a car in Phoenix and driving to a boat launch, where I met Brian to take a boat upriver to the base of a side canyon off the Colorado. We then hiked to the head of the canyon where the musicians and the crew had camped for the previous week. After a quick, simple dinner, we all headed farther up the narrow canyon, around a bend, and into a huge amphitheater of red rock, a circle of cliffs broken only by the narrow opening through which we had entered. I climbed away from where Paul would play. While the crew set up, I lay on my back, watching the darkening sky and listening to their talk about adjusting microphones and lighting the music that Paul had written that day. It was dark when he finally picked up his saxophone.

Paul was recording a series of lullabies inspired by the sounds and sights of the landscape—the canyon wren, the red cliffs, the thundering river—and written for his one-year-old daughter. It was slow, dreamy music, hypnotic in its repetition. It was the most beautiful music I had ever heard—the lone notes of the saxophone rising out of the canyon, the notes filled with the silence of the rock walls and the night sky studded with stars, with the coolness of the night air. He played for hours while I dozed in and out.

Brian eventually joined me. As we lay against that rock, beneath that sky, surrounded by those cliffs, and serenaded by that saxophone, I thought of all that drew me to Brian—our desire to nurture this quality of deep soulful connection with the earth, our commitment to doing what we could to create more places where this connection was possible, to hold onto those places where it was found. I remembered my respect for his vision and his ability to recognize what was needed and to get it done. Our time in Nepal—attending the Barun festival, living in Hedangna, trekking to the Tibetan border—and all the other journeys we had made came back to me, reminding me of the importance of time, of shared memories and experiences, time with my family, time with his. I eventually fell asleep, pressing myself against Brian and into the rock and trying to draw in any remaining warmth from the heat of the day, waking only when Brian nudged me gently, the light already beginning to appear in the eastern sky.

◊

For years, Brian and I had talked about whether we wanted children. I knew that I did, but I had to make sure that I had had enough of traveling and exploring to be content with staying behind. Brian was not so sure. He worried about the time children took and about having to give up the things that mattered to him and becoming trapped in a job to support a family. I finished my doctorate and began to teach, finally earning some money. Brian spent time with older friends who had strong, close connections with their children, and slowly he changed his mind.

◊

April 1998

I was in a hot tub. The only light was from candles. Julie's hand was on my sternum. Her mother had brought her from the hospice; she was talking about smooth white bones. Lida was holding a cup of ginger tea for me to drink. I was clutching the hand of Jane, the midwife; looking for some relief in her eyes; and repeating, "I can't do it. I can't do it—the pain is too much." She held my hand firmly and assured me that I could. I was doing it, she said. Brian was in the hot tub with me. Bear down on him, Jane instructed. Release into his support. Let him absorb some of the pain.

And then it passed, and I sat back and laughed at myself and said something about the music and that Julie looked like an angel, and then it came again, this wave of pain that was more intense than anything I had ever experienced. Jane told me to push, to push harder, but I could not. I was doing all I could. Release into Brian, she instructed again. Let him help. And I did, more than before. He did not flinch. I could do it, Jane said. I was doing it; it was happening, she repeated, again and again. And there was the head, the shoulders, the body, into Brian's arms and then onto my chest—a little spirit with dark eyes staring, smooth and wet against my skin. An opening. I opened. My body opened by her in this warm, watery room. Someone put a hat on her head. Jane helped me out of the tub and led me to the bed in the candlelit room next door. The baby, our baby, on my chest, eyes open. Julie and Brian and Lida stood close by. Jane stitched me up by flashlight.

Brian and I returned to our apartment in the early hours of morning, driving through the empty streets of the sleeping city. We stayed in the dream for days, Brian leaving home only occasionally to bring back food. In the evening, we lay together on the couch, Avery in my arms and both of us in Brian's, neither of us wanting to break the spell.

Every birth is a thin place, Jane told me a few days later, when we spoke on the phone. Many women do not understand that, and not all women can open to it the way you did, she said. I did not do anything, I replied. I always thought that opening required doing something. I simply responded to what was there.

◊

Time passed, and Brian returned to the world of deadlines and meetings—the things that led somewhere—while I stayed at home, by desire, doing the things I had never done: nursing and napping and changing diapers and nursing and picking up and nursing and changing diapers, putting beads on a string with no knot at the end.

Julie was living in a hospice, yet was remarkably well, given how critical her condition had been just two months earlier. More than anyone, she understood the experience of being between worlds, not here, not there. Even though she had not borne a child, she understood it in a way that many people did not. We shared the experience of being in a thin place, a sharing that helped it feel more comforting, less raw and exposed, the heightened awareness of living with nerve endings bare.

◊

As I spent time with Avery, I thought of all the details I had not asked about giving birth and raising children in Hedangna, practices I had not noticed because they were not related to my fieldwork. Except once, while walking up the hill with Pushpa and her two-year-old son, pausing every few steps while her son explored rocks and twigs, and listening to Pushpa talk about growing up in the village. And then she stopped, turned to me, and said that she had heard that in my country, old people were moved to separate homes where they lived on their own. "Was that really true?" she asked.

I had to nod and reply that it was true. Before I could explain the reasons or the exceptions, Pushpa spat in disgust. I said that sometimes that was what older people wanted as well, that it could be good, that it depended on so many things, but Pushpa was not listening. She said that she also had heard that children—even babies who had just been born—slept alone in dark rooms. Was that also true?

Again, I had to nod and reply that it was true. Again, I tried to explain that it did not mean that parents did not love their children, but was how my culture thought it was best to raise them, and that I had slept in a separate room as a child, as an infant, and being alone never bothered me. That is what I had known. And, again, she spat, not interested in any explanation I might offer to soften what I had said.

I read books, like every new mother, but only those that confirmed what I had seen in Hedangna: keep babies close to your body, bring them into your bed at night, let them nurse as much as they want whenever they want, create a container to extend the warmth of the womb.

◊

In the middle of the summer, Avery and I accompanied Brian to Shelburne Farms in Vermont, where he was helping at a conference about sustainability that was attended by well-known environmentalists. We arrived the evening before the conference began, to go to the reception. Avery started to cry, so I took her for a long walk in the dark, while Brian ate dinner with the others, talking about the environment and social change and what it took to make a difference, conversations I would have participated in and enjoyed, once upon a time. Avery finally asleep, I lingered outside in the shadows, watching the candlelight flicker across the faces of the men and women seated around the table, in a world somewhat less real than the world where I stood, wondering how to find my way back into the room, wondering whether I cared to.

I thought about Brian's and my worlds and how easy it once seemed to move between them. I had entered his world, in the past. And in Nepal, my world had intrigued him, so he did his best to cross over and understand what it meant for me, to experience it directly himself. Despite our physical closeness, it now seemed impossible to bridge our worlds. He came home at the end of a long day of work and spent an hour or two with Avery. In such a short time, it was hard for him to get beyond the foreignness, hard to create the closeness and comfort that I felt from being with her almost every hour of the day. Maintaining that closeness with and providing that comfort to Avery took all my attention. I had little energy for more.

◊

I spent hours watching patterns that became movement, as Avery tried to get her hand into her mouth, at first her thumb and then her fist, sometimes her hand pressed instead against her face. Sometimes she missed altogether, but then she got it and got it again. It made me sad, a new pattern gained, a bit more coordination, but something lost. She no longer bounced around a centimeter from my nipple when it slipped from her mouth, eyes shut, frantically hoping to bump into it again; now she was more assured, sliding the nipple into and out of her mouth, looking around, confident that she would get it back if it was lost.

My days shaped by this absolute focus on watching, watching more closely than I had ever watched, noting the movement of her eyes and the shape of her hands. Even when I took her on long walks, I carried her in a sling, running my hands along her bottom, cupping her feet in the palm of my hand, wrapping her fingers around mine.

My love for this little girl filled the spaces inside that were dry, like water seeping into a desert.

◊

I spent the weekend annoyed and lonesome, Brian absorbed in his work and me absorbed with Avery. I felt confused about the balance of work and about how quickly and easily we had fallen into roles I had not expected. The work we did each day—Brian outside the walls of our home, me inside those walls—drew us more deeply into our own worlds. I did not know how much of the distance between us came from the separateness of those worlds and how much from our relationship to them: my ambivalence about work and Brian's ambivalence about parenting. Whatever the reason, I thought that we would share more and that the magic of parenthood would overwhelm us both, would be a bridge and not a wall.

◊

Before Brian and I married, just before my first trip to Nepal right after college, Brian's parents bought the house and land in Canaan as a second home
in a part of the country that Brian loved. After a year or so, Brian and I were the
only ones who really used the house, and it began to feel like ours. It became
the landscape where we envisioned putting into place the dream that had
brought us together: living in the country, growing our own food, gathering
wood from the land, raising sheep, living more simply and with more intention. We had planned that after I finished my dissertation and Brian was ready
to leave or to change his job, we would move to that house and begin to live
that dream. And so in the fall, when Avery was four months old, we sold our
apartment in Cambridge, where we had lived for ten years, and moved into the
big, drafty house in Canaan, which we believed we could turn into a home.

◊

Walking alone in the woods above the house in Canaan, I felt the pull of my
other self, the self who lives in darkness in these woods. A shape among the
trees, ghosts that were more present, more real as ghosts than they would ever
be in the flesh. I felt that I was encountering another person—this other self
of mine—the one who wants to go out and away, who longs for the wind and
the rain and the wet.

How to live in a way where there was room for both?

◊

An appointment, finally, with Nora. She spent forty-five minutes placing her
hands on my arms and neck and back and legs, touching me as she had shown
me how to touch Julie after another session of chemotherapy, what Nora
called cellular holding, supporting the parts of Julie that were not cancerous.
Touch demands a kind of receptivity, she said, letting the fingers receive what
is there, a directness of communication that is not filtered through any inter-
pretation about what the feeling means. Bearing witness to another through
touch; opening to what is without turning it into anything else.

By running my hands over Avery's skin, Nora explained, I would give her a
sense of where she ends and the world begins, a sense of the boundary created
by her own body, not the one she would learn later with her head. As I touched
Avery's feet and shins and shoulders, she would begin to inhabit those parts of

herself, not just seeing the world but also coming to know her own presence in the world, the ways she does and does not take up space. I thought about this later. Those who do not fully trust their physical presence in the world see only the world, not their place in it. And those who know only themselves cannot encounter a world outside their own image. Touch is a means to support the movement between the two: receiving what is not ourselves, while expressing what is.

Brian once said that this book was about reaching my hands out and touching theirs—Devimaya's, Chute Rai's, Dhanmaya's, and Raj Kumar's—exploring what I could understand of their lives and what I could not, the differences being too great. I thought about the thin places where we let ourselves be transformed by touch, about the ways that these hours of touching, holding, and nursing this little girl are transforming me, slowing me down—no place else to go, no one else to be.

◊

In the late autumn, I packed Avery into the car, drove to Cambridge, and parked the car at the apartment where Julie had moved when she left the hospice. She then drove us to the airport. I waited in line to board the plane to Philadelphia, Avery on my hip, trying to read through my paper for the first time since printing it. Another woman with a child also was attending the annual anthropology meeting. Her husband was accompanying her, to help with child care. They were traveling a day early, she told me, to get settled. Her first panel was at midday the next day. She asked when mine was, glancing at the pages I was clutching. Two hours after my plane arrived, I replied. I mumbled something about having no time to talk and quickly turned back to my paper.

Several hours later, Avery slept in the sling against my chest while I stood at the front of a small, windowless room in a hotel and read my paper. I was on the panel "Memory and Home." I talked about how Raj Kumar, Amrit, and Ganga seemed so lost. I talked about how I had thought that villagers like Dhanmaya, who stayed in Hedangna, were simply those who could not get away. Leaving was an act. It was about being visible in the world, making a difference and being recognized for it, and constructing a story of your life. Staying, instead, was about being invisible, not taking charge, not being clear or creative or energetic enough to craft your own tale.

But I saw it differently now, saw the wisdom in Dhanmaya's assessment of her options, a wisdom that had to do with being able to distinguish between what could be changed—what could be controlled—and what could not. By understanding and accepting that her husband was not who she had imagined he would be and that her life was not what she had thought it would be when she was fifteen or twenty-two—by letting go of a story about who she was and where she was going—Dhanmaya could relinquish the idea of another life lived another way. She could come home to the life that she had.

The next day, I went on a walk with the two anthropology graduate students who had organized the panel. Both had done research in Nepal and were writing their dissertations. As we walked, we talked about the difference between *maiti* and *ghar*, the two Nepali words for "home." The *ghar* is a man's legal home. He retains rights to it until and unless those rights are taken away. He does not have to actively remember his home, even if he physically leaves it, because his claim as a son is indisputable. Once a woman marries and moves in with her husband or in-laws, her parents' home becomes her *maiti*, which means "love and nurturing," a place she has to remember because her memory of it is all that keeps it alive.

I thought about this division of houses, for married women but not for men. A woman is nurtured in one home, her *maiti*, by her parents, and in another, her husband's *ghar*, is expected to nurture others: children; soil; and, eventually, especially if her husband is the youngest son, parents-in-law. This distinction is not abstract. An older woman in Hedangna whom I did not know well once commented on how thin I had become. She asked if I had food of my own to eat between meals, if I got milk and ate eggs. I replied that I drank tea in my room, but not much more. "Don't be a daughter-in-law," this woman who was once a daughter-in-law and now a mother-in-law told me sternly. "Eat from your own hand, not from someone else's."

Although individual experiences varied, the legal system allowed men in Hedangna, and in much of rural Nepal, to stay whole in ways that women were divided. I did not think about this division when I was in Hedangna, so never asked men or women about it, only observing the consequences from afar. I could only guess the villagers' experiences from the perspective of my own, an internal division I encountered through the division of labor, not of property. As I watched everyone at the conference—talking, presenting papers, listening to talks—I thought about the divide between men and women in the United States, created and sustained by an economy in which women for the most part are the nurturers and men the producers, and by a culture that values and depends on production. I had anticipated this external division before becoming a mother. The more surprising split was the one I encountered daily within myself, between the self who thrived on being with Avery and the self who longed to start and finish things—from thinking a thought to writing a book—and to share those finished things with others. Handing myself over completely to the task of mothering drew me away from my earlier work. Doing that work, especially trying to do it well, pulled me from mothering. I was divided not because I was doing anything wrong but simply because I was doing what was ingrained in me to do.

I wondered why this realization came as surprise to me, wondered how my experience in becoming a mother was different from that of the women in Hedangna. They also bore the main responsibility for mothering, and yet the experience for them was an extension of their work as farmers in a marginal landscape; the villagers' lives—of both men and women—were structured around the day-to-day necessity of keeping their families alive. While mothering certainly added to that burden, made it more challenging for the women to meet those needs, having a child did not change the nature of the work that they had always done.

I had been raised in a world that was distanced from this edge of necessity, in which tending to household work—preparing food, washing dishes, sweeping the floor—felt like time away from the work of my mind, the work that I had been taught I was here to do. Yet suddenly, as a mother, I found myself immersed in a world where my main tasks each day were to meet the most basic needs, preparing food and changing diapers, and where on many days that was all I could manage. Given the vast differences in class, education, and opportunity between the women in Hedangna and me, it was absurd to compare my rights with theirs; there was no comparison. Almost all the women I knew in the village would have given anything for the comforts and physical ease of my life, for the experiences and opportunities I have had, would choose emotional rather than physical discomfort. And yet part of me could not help feeling that their culture better prepared them for the monumental task of mothering and that my education—which had taught me that a life subsumed with tending the basic needs of the body was somehow less valued than a life spent nurturing the mind—had failed to prepare me.

I attended the conference because I thought I should; it represented a world to which I should try to stay connected. But once there, I felt even more detached than I had in the past. As I student, I had felt alienated at conferences, always on the outside looking in. But then, at least, I had known who was who and what the important theories were. I had known which panels would be interesting to attend and actually had attended them. I had been on the outside but cared about moving inside. Now, I realized that I did not even know which talks I should try to hear. Sitting cross-legged in the carpeted hallway between the rooms where the papers were presented, I found myself utterly absorbed in Avery's delight as she realized that she could move herself across the floor, having learned to crawl the night before we had come. I had no idea where this little spirit would lead me, and yet I knew enough about the journey with her so far to know that this was the one I wanted. What was opening inside was what I had longed for—the same longing that had taken me to Nepal. I was struck by the irony of having spent all those years in graduate school, ostensibly to be able to converse with the people inside these rooms, and now discovering that the thread that had taken me to Nepal was what kept me outside these rooms, following my daughter through the carpeted hallway.

◊

One snowy winter night when Brian was in New York City, Avery was sick, hot with fever, tossing and turning and sobbing for what seemed like hours. It had already snowed two feet, and another foot was predicted. In the middle of the night, I finally wrapped her in a blanket, climbed out of bed, and carried her to the window, where we stood in the cold, dark room and watched the headlights of the snowplow that finally was making its way up our long driveway. We watched the lights go back and forth through the freshly fallen snow, as the plow cleared a space for the cars. When it finally turned to head back down the driveway, Avery watched the shadows slide across the faded wallpaper of the bedroom, following the movement of light on dark, outside and then inside, not, as I had done, watching the plow finish its work, the completion of what my mind determined to be the action. For her, the gliding of the shadows was as substantial and significant as the plowing of the snow.

The next night, I took a sauna in a converted icehouse in the woods. I lay on a shelf near the roof and looked through the window at a bare, spare birch lit by the full moon against the deep midnight sky and at the smoke rising from the chimney. I tried to imagine what the world looked like through Avery's eyes, what she saw when she looked at the lights and snow, receiving the images without immediately identifying them—tree, moon, smoke—trying to see as Nora touched, letting bone and muscle become present on fingertips. In the moment I could see that way, the world became enchanted. Everything seemed alive because I was seeing the world as it was, not its labels, nothing to distinguish between a tree moving because of the wind and a tree moving because something inside felt like moving.

◊

An opening inside slammed shut when I learned that on the feverish, sleep-less night on which I watched the snowplow with Avery, Brian had been having dinner with another woman—whom I also knew—in New York. I had not known that they would be in the city at the same time. Brian said it was nothing. He saw my expression, and told me again that I did not have to worry. But I did worry. The world pressed in and threatened to make me a margin to someone else's center, make me forget what I was learning to see and know on my own.

◊

Brian and I climbed Mount Cardigan in the snow, Avery asleep in the back-pack, and talked about the balance between responsibility and longing, the ties that bind and those that release. The dinner in New York confirmed an unspo-ken tension between us. We talked about the fear of setting out on a journey, about the risk of getting lost physically and psychically, and about the equally fearsome risk of losing your soul if you did not go. He spoke about opening to the mystery, the fear of losing a part of himself if he did not follow where he was being led. I understood this fear. I did not want that, for me or for him. That is what had continued to take me to Nepal. I said this, but we talked about it in the abstract. Although Brian reiterated his commitment to make our rela-tionship work, I doubted that he was so sure. Our home had become a burden. Our relationship needed work. Life with a ten-month-old was not easy, espe-cially since we were so physically, professionally, and socially isolated. Nei-ther of us mentioned it, but I felt it every night when we sat down to another dinner, Avery in the highchair and the two of us at the table squeezed between the sink and the refrigerator, and knew it in the sameness—a sameness that had become oppressive—of our days, both of us working from home, earnest in our effort but without a lot of joy.

I thought that falling in love with another person, devastating and confus-ing as it could be, did not have to be a reason to end a marriage. To me, it could be an opportunity to be more honest and open with each other, to show up for our relationship in a way that neither of us had in the past, and to create the intimacy that both of us longed for. It seemed that leaving was the easy way out; at least, it felt that way to me because the work of staying seemed so daunting. It seemed that leaving was giving up on a journey we had really just begun. But I had left emotionally as well, although had chosen to return. Now, as a mother, I was willing to do what I could to transform our relationship into something real. Not just because I did not want Avery to be split between two homes, but because of what I had experienced in Hedangna and what I was coming to know more fully through Avery, seeing that whether the world was enchanted or not depended on me—on my own presence of mind, body, and spirit—as much as on the world.

I said some of this as well, but did not say it all. I spoke from an experience that I did not share, adding one more layer to all that kept us apart. And I hid a fear that I did not want to admit, even to myself—the fear of becoming invisible, which I was becoming, had become, to Brian, even as I was becoming visible, finally, to myself.

◊

I thought how easy it had been to be the adventurous anthropologist traveling around the world, easy to be the sort of woman Brian was now drawn to. A part of me I once knew so well listened from a distance as friends talked of returning to Nepal and India, of hiking and backcountry skiing. Watching Brian's anxiety and excitement about his projects, I was frightened by the distance between us, yet I could not find the part of me that cared to close the divide. I remembered Dhanmaya lifting a basket of rice to sell at the market in order to give Devimaya money for her trip to Khandbari, even as she knew how hard it would be to fill the hole created by her daughter's absence. I thought of the part of me that could be in the background, the part whose comfort and voice could be set aside—with the confidence to set them aside—trusting not just that I would be able to come forward, when it again was my turn, but also that there was something to be learned in the shadows. This assurance was new, sensing, for the first time, perhaps, a kind of feminine power: moving with, not pushing through.

I always thought that mothering would require denying part of myself. I now saw it as a way to inhabit a part of my body, mind, and soul that I could inhabit only when I loved without the need to possess, when the act of loving was the gift I received.

◊

Walking through the woods with a sleeping Avery, I thought about mothers and shamans and why no women are shamans in Hedangna. Being a shaman requires seeing that some things are not meant to last, while being a mother means working to make things last no matter what, about fiercely holding on.

Is it possible to embrace both? To be attached to a world you are also willing to leave?

I thought about mothering and farming the way the villagers farmed in Hedangna. Their intimate connection with the soil allowed them to respond intuitively to it, the way I responded to Avery, anticipating her needs as if they were an extension of my own. This quality of intimacy is accessible not only through mothering, although I came to know it as a mother. It is an intimacy that made porous my own boundaries, seeing that Avery was in me and I in her in ways I could not even begin to understand—a relationship that took time and attention, in which the things that mattered were the things that could not be seen. What does responsibility—the ability to respond—mean, then? What does it take to pause before acting, to remember that the things that seem to be outside, the people and places we love, are also inside, shaping what we see and do in the world? Then limiting our desires is not about denying ourselves, but about coming more deeply into ourselves, as we open to all the selves inside us.

I thought about the different kinds of love, the love of a lover, of romance where the beloved is the star, love where we give and we receive—love that is visible. And there is the quieter, more solitary love of a parent for a child, like growing rice. No shining light. The invisible daily work.

◊

Lying in our bed, I tried to nurse Avery to sleep, but she nursed for a while, drifted to sleep, and then startled herself awake, and I had to start all over. I patted her back and sang in her ear. She kept pushing up on her forearms and crying, then dropping down, pushing up, and then dropping down. Up and down, up and down. I wanted to go downstairs to read and talk like an adult, yet nothing I did seemed to work; my frustration became rage, which became despair, keeping us locked in a struggle that kept sleep at bay.

Finally I gave up resisting her resistance to sleep. I instead lay next to her and told her the story of my journey to Khembalung, climbing the cliffs to the caves. Both of us dozed off, and I dreamed of a woman—tall and slim, with long dark hair—walking across a barren, rocky landscape, bare feet on sharp stones, toward the headwaters of the Barun, silk scarves flowing behind her in the wind as she walked across the land, leaving everything behind. She took off her dress and scarves and dove naked into the icy mountain lake.

The next day, instead of doing yoga in the cold dark room, I danced. Streams of light seemed to flow from my hands and my hair. As I danced, I was the tall, slim woman with long dark hair and flowing silk scarves. And as that woman, I could stay.

◊

Shortly after Avery's first birthday, Brian told me that my wings had been clipped by motherhood. We were fighting at the time, and perhaps he said this more for the effect then because he meant it. But I was stunned that he would say that, stunned at the distance between us that allowed him to think that. I had lost the wide-angle vision I once had, but in my absorption in the day-to-day life of an infant, in my focus on her immediate needs, I was seeing and feeling things I had never seen and felt before, things that would inform everything I saw in the future. I was shocked that Brian could not see that, could see only the ways I was not moving, not the ways I was.

I once thought of invisibility as an absence, an emptiness. Now that the qualities I was coming to value most in myself had been rendered invisible, I understood that invisibility was simply the presence of things that were not readily visible to our eyes, not because they were not present, but because they were no longer valued by the world in which we lived.

Once a month, women in each household in Hedangna gathered a pie of cow dung from the fields. They carried it home, where they mixed it with red mud gathered from the hills and water taken from the spring. They made a thick paste, which they rubbed over the inside walls of the stone-and-mud houses. There seemed to be little to show for this work. Only when it was not done did it become apparent; the walls began to dry and crack, and eventually they fell down. As I nursed Avery, bathed her, read to her, sang to her, and tried to get her, again, to drift off to sleep, I thought of the men and women in Hedangna and of all the work needed to produce enough to feed their family for one more day. And as I changed her diaper or prepared her food, I imagined women's hands rubbing mud over the walls of the houses: the ongoing, invisible work needed to stay in place.

21

SAGE MOUNTAIN

April 1999

The summer I became pregnant with Avery, I attended a women's medicinal-herb conference in southern New Hampshire. Three hundred women gathered for the weekend to learn to make tinctures, teas, and salves; to study the medicinal uses of common weeds; and to identify those plants in the wild. I had been intrigued by what I had learned about medicinal plants from an herbalist friend, and I wanted to learn more. I went to as many classes as I could, taking pages of notes about which herb to take for stomach ailments or dry coughs or wet coughs. But what I mostly remember from my first encounter with traditional herbalism was the women who practiced it. The way they spoke about their lives and the plants, the way they interacted with the children who were there, and, mostly perhaps, their own childlike wonder seemed to express a coherence between their selves as mothers and their selves as herbalists that I had never seen before. Nurturing and nourishment was their way of being in the world, their way of caring for themselves and others as they produced the things needed to stay alive, not an additional task to do if they had any energy left at the end of the day.

Just as being in a sacred place opened me to the sacred parts of myself, being in the presence of a wholeness I longed for helped me feel whole myself. And so, when Avery turned one, I enrolled in a weekend apprentice program in the art and science of traditional herbalism at Sage Mountain, in central Vermont. For the past year, I had experimented with making different teas from plants in my garden, but all I really knew about plants was what I had learned in a course on tropical ecology in graduate school. I chose the program at Sage Mountain because it was the heart of traditional herbalism in New England and because it was only an hour from Canaan. Mostly, I chose it because I wanted to study with Rosemary Gladstar, one of the leading herbalists in the country and one of the two organizers of the women's medicinal-herb conference I had attended. I had been especially impressed with how she talked about plants, weaving together science and spirit in a way that seemed so fluid and natural. I wanted to learn more about how she brought a sense of the sacredness of plants into teaching about their uses as medicine.

◊

Sage Mountain was unlike any other place I had found in New England, a clearing carved out of the wilderness. It was wild, but not only wild. The first morning of the program, I pitched my tent in a meadow and then walked through the woods to a canvas yurt set in a small clearing. It was late spring in Canaan, and thus still early spring on the mountain. I wore a winter coat. Thirty apprentices—twenty-eight women and two men—sat in a circle in the yurt and listened as Rosemary, wavy auburn hair hanging down the back of her long, maroon velvet dress, talked about health as wholeness and healing as a process of restoring balance. She described how medicinal plants strengthen the immune system, nourishing and supporting the body's own capacity to heal itself, trusting the integrity of the cells to decide what to assimilate and what to eliminate.

What Rosemary said reminded me of the Yamphu concept of *lawa* (spirit or soul). Like the boundaries around *kipat*, where the security of a land claim depended more on the relationship between neighbors than on rigid demarcations, the boundaries of the body were porous. There was nothing the villagers could do about that. The ways to prevent a soul from being lost were to strengthen the health of the soul itself, giving it rest and good food, and to make offerings to the ancestors so the world outside was less dangerous, not by reinforcing the boundary between inside and out. We become and stay healthy, Rosemary was saying, not by keeping things out or depending on others to keep them out, but by fortifying ourselves from within and working to make the world a healthier place.

◊

After lunch, Rosemary sent us into the woods to sit with a plant, whatever plant we were drawn to, and ask what medicine it had to offer. At the conference, I had been most intrigued by and most skeptical about many of the herbalists' accounts of having learned from plants themselves. Some had taken their stories too far for me, talking about *devas* and fairies, but what others said about the importance of connecting with the life force of the plant and learning intuitively from that spirit resonated with what I had learned in Nepal. Wary as I was, I was curious to learn more. And so I followed the others out the door, choosing to climb the hill behind the yurt into the forest, which had been severely damaged from an ice storm the previous winter. The tops of the trees had fallen, letting the sunlight reach a forest floor thick with ferns and saplings growing in the light that now shone directly where once it was filtered by branches and leaves. The place felt as mysterious as it did familiar, either from the luminescence of the pale-green ferns or from a sense of presence, a coming together of the sacred and the profane in a way I had experienced only in Nepal. The damage from the storm had created something beautiful, a brokenness in the land that drew out and made safe the brokenness I felt within. It felt like a stone resting platform in Nepal, where I could lay down my burden and rest before picking it up and continuing on my way.

I sat next to a plant I did not know. I felt silly, but there was nothing else to do, so I started to talk out loud, as Rosemary had suggested. I said that I did not know what was unfolding, who I was becoming, and who Brian and I were becoming and that I had come in search of a way to feel less alone. I then sat quietly for a while, looking down at the seedling. It was seven or eight inches tall, with intricately patterned leaves—like a network of green rivers and streams seen from above—already unfurling. The forest floor was covered with these plants, whose leaves were so much more elaborate and delicate than the slender, oval leaves of the bellwort nearby. I was struck by that difference, by the extravagance of the leaves; they seemed so bold and so unnecessary. I wondered what the plant was, what its flower would be like. I closed my eyes, remembering that it was early spring, too early yet to bloom, and sensing at once that the lesson I had to learn was to be where I was now—on this mountain and by this plant. And I had to be myself, no matter what every other person around me was doing. I had to become the plant unfolding inside me, to let that be my meditation.

I walked back down the hill with a sense of spaciousness that I had not felt for months, a feeling of distance between me and the tension at home, a pause in which to catch my breath, to watch the steam rising from the bowl. The only thing I can control is whether I am in this moment or not.

◊

After the first weekend at Sage Mountain, I began to spend time getting to know the plants I could find in the meadows and woods near our home: nettles, burdock, dandelion, red clover, calendula, blue vervain, and St. John's wort. I gathered and planted them and made teas, tinctures, and salves. This pause in the busyness of my life allowed me to connect with the source of the medicine I prepared for my family, especially for Avery. Making my own medicine was as deeply satisfying as growing my own food. Rosemary talked about some of this, about how healing comes from removing the conditions that create disease on the inside and the outside, from tending to our minds, bodies, and spirits as we also work to live lightly on the earth. It made such sense to me that this attention to wholeness from the inside out might offer a path to help heal the wounds on our bodies and the earth inflicted by a culture and an economy that created divisions.

◊

Earlier that year, Brian and I had begun to plan a four-month trip around the country the following autumn. We would gather information for a newly conceived book project, again envisioned by Brian, that explored concrete examples of the relationship among land conservation, culture, and community in the United States. We planned to visit ten conservation projects in different parts of the country that best expressed the themes we intended to focus on: pilgrimage, good work, forbearance, memory, belonging, home, and others. Our intention was to find stories that I would document with words and Brian with photographs, to illustrate how the experience of protecting land filled a spiritual vacuum in contemporary American society. We were looking for projects that had created a positive change in communities: greater civic engagement, more creativity in community life, stronger local economies. We hoped to find projects that had helped communities adopt land-based values and, especially, that told a good story with interesting individuals and powerful, clear outcomes.

This project was part of a larger effort that Brian was involved in to document the qualitative impacts of land conservation. At the time, much of the value and importance of conservation was measured in terms of acres protected and money raised. After ten years of working within that framework, Brian wanted to move beyond dollars and acres to the cultural dimensions of land protection, to describe not only how it was saved but, more important, why saving it mattered. I was also interested in the cultural dimensions of land conservation, especially the cultural changes that conservation brought about. That shared interest had led us to decide to work on the project together.

We planned to gather this information in communities stretching from the northeastern tip of coastal Maine to the Hawaiian Islands and to complete this phase of the project in four months. The time seemed so short—as an anthropologist, I had spent almost two years in one village of 270 households—and I worried what we would miss by not spending more time in the communities whose stories we were telling. But I had specifically chosen to work on this project with Brian because I wanted to collect stories that changed people. So often I felt paralyzed by the feeling that nothing made a difference, that we were all trapped in structures and institutions, all with their own rules, their own logic, inextricably tied to a worldview and culture that was paving over those moments and places where life endured. I thought it would empower me and others to make visible some of the less tangible benefits of protecting land and help keep my despair at bay. Given our intention of visiting diverse communities that were struggling with a variety of issues, it was not feasible to spend more time than we planned.

We were also taking this trip to save our marriage. This was one of our reasons for deciding on a four-month journey instead of a series of two-week trips around the country. We traveled well together. I hoped that the excitement of sharing ideas, which had brought us closer in Nepal, would spill over and rekindle our relationship.

This is what my head told me. But I was feeling more and more uncertain that it was the right thing to do. During the months leading up to the trip, the only time we seemed to get along was when we were making plans for it. Whenever I expressed doubts about Brian's commitment to our relationship and about whether going away for four months was the best idea at this time,

Brian dismissed them, pointing to all the attention and time he was devoting to the upcoming trip. He did not have to undertake this project, he said. But he was. What better sign of his commitment did I need?

◊

July 1999

In a sweat lodge at Sage Mountain, I called on Chute Rai and Baiseti Thuma; I called them in Yakabha, *"Thuba, Thuma: Grandfather, Grandmother."* I knew the risk for anthropologists of going native. I was not trying to go native; I was not running from my past or ignoring who I was, but simply trying to come home in a way that honored and built on what I had discovered. And I needed them to remind me.

I called their names again: Chute Rai, Baiseti Thuma. I was calling them now, I said, because I did not know my own ancestors. I did not know who else to call, so I was turning to them again, asking them to pull down a straw mat and talk to me, to create a moment outside regular space and time where I did not feel so alone, where the distance between me and the world did not cut so deep. I asked them to help me come home—home to myself—so my homesickness would go away.

I then sat in the darkness and the heat and realized that even as I knew this was what I had to do, I still longed to come home to a person and a place outside myself where I would feel held and safe and whole. That was what Brian and I had hoped to create in Canaan, the home we had imagined: a house in the country with a large vegetable garden and flowers, sheep, and woods in which to wander; a child or two; and work that would make a difference in the world. All the pieces were in place, but something was missing, something that mattered more than any of the rest. Without that something—call it the essence in life that makes the food last, that makes us all last—all we had was an empty shell. Coming home, I finally began to understand, involved filling that shell with life and light, imbuing it with a presence, living life as a pilgrimage, at home and away.

Speaking into the darkness and the heat, I said that I had come to a place in the woods where the trail ran out, where it was dark, and I did not know where to go. Then I listened as the leader of the sweat talked about the spirit world and an inner healer. And then it was silent and hot and dark, and I kept longing for a vision, for a guide to appear to lead me on a journey that would reveal the healer inside. But nothing happened. And then during the fourth round, I stopped trying. I opened my eyes in the blackness and saw the coals glowing and heard the chanting and felt the presence and warmth of all the people I had not noticed before. I opened to what was here and discovered that I did not need anymore. This here was enough.

22

SACRED STORIES

September 1999

One sunny morning at the beginning of September, we loaded our new green Volkswagen van with stacks of books on land and people, camera bags, two laptop computers, shorts, long pants, warm sweaters, and sleeping bags—everything we could possibly need on our journey across the country. Kate, a student in a class I had taught on culture and the environment the previous summer, had arrived a few days earlier; she was coming with us to help watch Avery and to learn from the communities we visited. Now that we were on the verge of leaving, the trip seemed daunting. I wondered if it would be too much movement for a sixteen-month-old: too many different people, too many different places. Avery was a difficult sleeper even when we were in the same place every night, and she hated her car seat. But it was also an adventure, and, especially after the past year of feeling so trapped in place, it was exciting to be heading out on the road.

The night before we left, we performed a simple ritual by the garden. It was the night of the new moon, so I filled a glass vase with calendula, mint, and lemon balm from my garden. We placed the vase in a circle of stones to leave overnight so the infusion could steep in the darkness of the moonless sky. The next morning, we would drink the tea in honor of new beginnings.

After making the tea, we gathered on the grass in a circle in the late-summer dusk and talked about our hopes for the journey. After we spoke, we offered objects that expressed those hopes, which Kate later placed in a small cloth bag that hung from the rearview mirror. Brian went first. As he spoke about his vision, about land and people, I realized that I was not interested in helping create a renewed land ethic in America by documenting the cultural dimensions of conservation. If I was honest, I was not sure that I cared anymore about discovering how other men and women had come home to the land. I realized that I was embarking on this journey—an American pilgrimage of sorts—to try to find my own way home. Only when I had done that could I decide what to do with my marriage. I had to come home completely before I could decide whether to leave.

An anthropologist in Nepal once told me that the witches and shamans make deals with each other. (Witches tend to be women in Nepal, and shamans typically are men.) A witch snatches a person's soul, the person becomes ill, and a shaman is called to recover the lost soul. In exchange for the knowledge needed to find the soul, the shaman must offer the witch something he holds very dear. The more he wants the information, the more valuable the offering must be.

For me, the stakes were high. Brian and I had been together for seventeen years and married for thirteen. We had a sixteen-month-old daughter. What we lacked in emotional intimacy we made up for in the closeness of a shared life and shared memories. I could not imagine walking away from that, and I could not believe that Brian could.

And so, like a shaman, I offered something that meant a great deal to me. I put a piece of green glass inscribed with the word HEALING into the bag. Julie had given it to me the last time I had seen her. I had refused it until she showed me another piece of glass, also given to her by her mother, a blue one etched with the word CURE. It was just a bit of glass, but it reminded me of the openness of Julie's heart, which, in turn, reminded me to open my own.

◊

Early the next morning, we headed to Helen and Scott Nearing's homestead on the coast of Maine. We had selected the Nearings' house as the place to begin our journey because, as with sacred mountains and temples in far-away lands, people traveled from all over the country to visit their handcrafted stone home and organic garden and to experience directly their vision of the "good life." We thought that it would be a perfect place to talk about pilgrimage and sacred places.

We stayed in their original homestead, a small white farmhouse up the hill from Forest Farm, where they had lived in their later years. A stone wall still surrounded their original garden, which, like much of the land, was now overgrown with weeds. A narrow path through the woods connected the homestead with the more public and busy Forest Farm.

As we spent time over the next few days with neighbors of the Nearings and with visitors, I was again struck by the gap between the image the Nearings portrayed of their life and the reality of that life: things that the Nearings had chosen not to mention in their books. To me, these things—earning income from sources other than what they produced on their own land, hiring someone to build the stone house that they claimed to have built themselves—cracked open the perfect vision of self-sufficiency. I thought back to the chicken feathers swirling against my windshield, to the tenacity and clarity with which the Nearings had pursued their vision. People visited the Nearings' home and garden because they were inspired by the life their books described and by the clarity and power of the vision that had guided that life. We were at Forest Farm to tell the story of how those visitors were transformed by their encounter with the life and home of the Nearings. That story, we hoped, would motivate them and others to incorporate parts of the Nearings' vision into their own lives: to live more simply and closely to the earth. How could we—should we—accommodate this information, which indicated that the Nearings' life on their land was more complicated than they had presented it to others?

Stories are one way—perhaps the most important way—we make sense of our lives. They are what we tell ourselves and others about who we are and what we want, about where we are going and why. They provide a fixed point, a secure place from which we can step into the world.

I knew why stories matter. I knew their power. And I knew the importance of a clear vision. But now I was thinking about Nepal. Like the myths of indigenous people and of the modern world, the stories the Nearings told created expectations about what they had accomplished that could never be fulfilled and that perpetuated the disease their books claimed could be cured by going back to the land. Like those myths, the Nearings talked about the good life as being defined by a set of actions—growing food, building a home— that could be measured. If people lived in this way, their lives would be good. Their stories felt like mirrors, reflecting all they had accomplished, not windows into something larger.

We had come to the Nearings' home to write about pilgrimages, not to explore the silences and contradictions. But to me, there was a connection. Pilgrimages are about stripping away, digging beneath the surface to something deeper that endures. I thought of my journey to the Khembalung caves, of how the pilgrimage itself called forth qualities in me, an opening and an attentiveness—walking, walking, watching, watching—that, in turn, allowed me to connect with something beyond myself—call it the sacred.

I thought about the difference between the Nearings' stories and the *mindhum*: Yamphu sacred stories about the founding of Hedangna, about the ancestors—where they came from and what they were like—and about passing through the *tsawa* and entering the realm of spirit. These tales, like a pilgrimage across the land, are a peeling away, a path to a clearing in the woods. They are about going deeper into life—a pilgrimage of the soul—connecting the listeners to the invisible world that sustains them and reminding them that they are part of a larger living and more-than-living community. They are not about defending territory, any sort of territory, and so there is nothing to hide. They are about seeing in a different way, not about being seen or making a claim. Listening to them, I noticed things that I had not before; I slowed down, paying more attention to the words and the meanings they depicted. There was no place to go but deeper into the story, deeper into the moment— like entering a clearing in the woods.

The Nearings' stories are inspiring—growing their own food, building stone walls, having time for music and reading—but they also cast a long shadow of perfection. Reading *The Good Life*, I felt inadequate. The book made me think about all the things in my life that had to change, all the choices I had made that were wrong. Instead of feeling connected to a larger community, I felt overwhelmed by all I had to do in order to connect with those who were really living the good life. And to the extent that I felt I could connect with that community, I feared that others would discover the parts of my life that were less "good," those choices I had made that did not pass the test.

Like the stories of the Nearings, the stories that Brian and I told each other about what we hoped for on the trip and in our marriage created more distance, not less. Each of us talked about what we needed from the other: Brian needed trust and respect from me; I needed Brian to listen and to try to understand my fear. Each of us was so focused on telling our own story that neither of us could pause to really comprehend what the other was asking. Instead of revelations, our stories became versions of the world that we had to defend, like a plot of land: our last line of defense against the unknown.

◊

Several days after arriving at the Nearings' homestead, knowing that I should not do it, I opened the neatly closed zippers of Brian's camera bag. It was the one thing he kept with him at most times, the one place he would hide something private. I knew that my searching and questioning were driving him away and that I had to trust my intuition about whether he was telling me the truth. But I did not know if my doubts came from my fears or from sensing something in his actions and his silence. I needed confirmation. I unzipped each pocket, heart beating, hands shaking. I kept telling myself to stop. I kept unzipping pockets. I did not find anything. When I had almost begun to believe that my suspicions were unfounded, I opened a small outside pocket. Inside was an intricately carved bamboo tube. Feeling numb, I twisted off the lid and slowly pulled out a piece of paper coiled inside. The paper was handmade. I unrolled it and found a handwritten poem with tiny sketches in black ink along the margins. The handwriting was that of the woman he had assured me he had come on this trip to forget.

◊

I believed that immersion in a tradition or story, both feet inside the circle, would close the distance between me and the world around me, a distance that I thought had been created and perpetuated by my training as an anthropologist, a discipline that encourages participation while always keeping one foot on the outside. And so I wanted to believe the Nearings' version of their life; more than anything, I wanted to believe that some people could make it work, that the pure far-off places still existed, even if the *tsawa* had been filled with a block of cement. I wanted to believe Brian. My inability to accept these stories at face value, and yet my desire to do so, created a tension within myself and with Brian that felt irreconcilable. Intellectually, I could accept the idea of treating stories as pilgrimages, peeling away the layers to the kernel beneath. Personally, I realized that it was much more difficult.

As I pulled the bamboo tube from the camera bag, I realized that there are reasons to cling to the stories we tell one another and to hold onto the categories that prevent us from seeing the truth. It was not only that I was naive. We need tools to respond to what we discover when we pay attention to others' gestures as well as their words, to what our bodies know, beyond what we want or think it should know. We have to know what to trust when the story we have believed in, the story we have told ourselves and others—the story we have built our life around—comes to an end.

I knew that I could not erase the past by putting it back in the pocket and zipping the pocket shut. Yet I also knew that taking out the bamboo tube—or Robert's e-mails—and looking honestly at what it meant, for Brian, for me, for our relationship, took more courage and skill than either of us had at the time. Brian explained the poem away. I let myself accept his explanation. But before leaving the Nearings' homestead, I removed the HEALING glass from the cloth bag hanging from the rearview mirror. Without telling the others, I slipped it into my pocket. I was not ready to let go of the story of our life. But I also was no longer able to believe it.

23

LISTENING

September 1999

After leaving Scott and Helen Nearings' Forest Farm, we headed west, stopping for the night at Kate's parents' house outside Madison, Wisconsin. We had been in the van all day, driving past shopping malls and 10,000-square-foot houses eating up the Wisconsin farmland. We were feeling grumpy. At the house, we took turns watching Avery so each of us could go for a run.

I ran down a dirt road and then turned up a trail that flanked fields of alfalfa and corn, running more to escape my despair than for exercise. I did not pay attention to where I was going, and the trail ran out in a field filled with rows and rows of alfalfa: the agricultural version of the landscape we had spent the day traveling through. With relief, I saw clumps of burdock growing along the edges of the field beyond the reach of the herbicides sprayed to keep them out of the alfalfa. I squatted next to one, looking closely at the burrs and the huge green leaves. After a few minutes, I closed my eyes. I saw my truck in

Maine that hot summer day, tar fumes coming in the open window, and then an image of the Nearings in their garden, day after day. I felt a sense of what it takes to stand one's ground on the margins. I thought about how hard it is to dig up burdock roots and how healing they are: the healing, perhaps, that comes from that kind of tenacity, from upholding one's own vision in a world that demands something else.

In Hedangna, I had been surprised to discover that the chants I transcribed from my interviews with Kelekpa and Purnamba were mostly lists of the names of flowers, other plants, trees, and places. The chants described the journeys and experiences of the ancestors, but depicted them in the names of the things the healers passed on their travels, not in what they did in those places. I had not known what to make of these lists and focused instead on the sections where something seemed to happen. It was only years later that I realized what I had missed in Hedangna. Perhaps each plant named in the chants told a story, like burdock. I was looking for stories in a form I recognized and so missed out on the ones I did not.

◊

The next day, we headed north to the Red Lake Reservation in northern Minnesota. A land conservation organization had helped the Red Lake band of the Ojibwe buy land that was farmed with wild rice. We had chosen this project as a way to talk about the reemergence of indigenous ways of farming and, more generally, about sustainability and right livelihood. During a meeting on the first morning of our visit, the members of the tribal counsel explained that the project was significant because it represented the first time Native Americans had gotten land back from whites instead of losing it to them. They did not mention anything about sustainability.

The reservation stretched along the southern side of Red Lake, the sixth largest lake in the country, a huge expanse of water in which, because of overfishing, fishing was prohibited. Not one boat was on the water the whole time we were there. The emptiness of the lake was eerie because it was so unfamiliar. Scrub woods thick with brush and fallen trees, broken only by dirt driveways, grew close to the edge of the road. There were no open meadows or fields, no lawns. The land, like the community, appeared difficult to enter.

We were not sure what to say. We had come to Red Lake to find an example of traditional ways of working the land and instead discovered wild-rice farming that was not traditional, work performed by white men who received their salaries from the Indians. As with the Makalu-Barun Conservation Project in

Nepal, what mattered was who owned the land. It was only after ownership had been fully established that conservation and sustainability become topics the people felt they had the option to consider.

We did not have the time to do the work needed to uncover the story that was really here, to get beyond our objectives and begin to understand those of the community. That was not why we had come. Now that I found myself in a similar position, I realized how quickly I had judged Khagendra and the Makalu-Barun Conservation Project for not doing what I, here at Red Lake, was also unable to do. I again thought about William Cronon's comment about the black box and how quickly I had nodded, imagining that what he advised would be easy.

◊

What I mostly remember of our stay at Red Lake was sitting around a linoleum table with tired gray nuns, all over sixty, eating overcooked potatoes and cabbage and boiled meat. The food was flat, like the land. I envied Kate, who spent her mornings wandering in the scrubby woods with Avery while we tried to interview members of the tribal council, tribal elders, and employees of the federal government who worked for the tribe. I could not get beyond the stories of alcoholism and poverty, of broken families and abuse. "We tell the white people they can leave off hurting us for a while," the head of the tribal archives told me. "These days, we're doing enough harm to ourselves on our own."

The culture and the history of this reservation was about keeping outsiders away. We had hoped to tell a different story that would make people who read it feel positive about being connected to the land. What we discovered was more complicated and difficult than we had naively expected, tied to a history that was impossible to ignore—a history that was not ours to tell.

◊

The difference between Hedangna and Red Lake—was it a difference?—was that in Hedangna I had had time, time to get to know people before and after asking questions, time to give something in exchange for the stories I was taking, time to listen for what was not said along with what was. This time had allowed me to understand more of the connections the villagers made, the meanings they drew from events, their reasons for what they did and the choices they made. But ultimately, I still had interpreted their stories through my experiences. I had used the information I collected for my own purposes, just as we were using the stories we gathered on this trip for our own. When is recording stories a way of enhancing and deepening cross-cultural communication, and when is it yet another form of cultural exploitation? When is it appropriate to try to close the distance between cultures, and when is it best to let it be?

◊

Brian, Avery, and I slept on cots in a room that had been a dormitory on the top floor of a large building that was also the convent where the nuns lived. The carpets were as worn as the furniture. This building had housed the boarding school on the reservation, one of the most culturally devastating institutions in Native American history. Children had been taken from their homes and reservations and sent to such schools, where Anglo teachers tried to force them to abandon their culture, language, and religious beliefs in exchange for English and Christianity.

The second night I dreamed of a child defecating on the floor by mistake; I saw feces smeared across the wall and the child being beaten. I heard voices shouting and young girls crying. I awoke and lay on the narrow bed, staring down the long, dark room lined with empty cots.

In Hedangna, when a number of children die in a house, a shaman is called to determine whether a curse has been placed on the land or the house and, if so, to perform a ritual to remove it. Sometimes, the ceremony works. Other times, the curse is so strong that the family has to build a new house on a different plot of land. Aldo Leopold wrote of how "the sadness discernible in some marshes arises, perhaps, from their once having harbored cranes" (1987:97).

◊

Starting at Red Lake, and continuing through our trip, I had the recurring feeling that there was something in the background, a story behind the story that was trying to emerge but could not be captured in formal interviews and ten-day visits. But you need resources to follow different threads: money and energy and time. Our clarity and vision had raised the money for the trip, whose goal was to find particular stories that spoke to particular themes decided more or less in advance. We could not accomplish this objective if we followed every thread we encountered, and the visions that had come to me as I crouched next to the burdock were not likely to impress many people or inspire them to act. Even so, I longed for a model of storytelling that was open-ended, not imposed, as was Khagendra's map, by one group on another.

Louise Erdrich once began a talk by noting that in his journal Christopher Columbus wrote about his desire to take one or two natives to Spain to teach them language. He wanted to teach them language, she repeated. Columbus did not pause long enough to consider that the sounds the indigenous people were uttering were words, a language—their language—that served their culture the way his language served his own.

She paused and then continued, "I want to talk about what it takes to listen, what it takes to stop the static of what we already know, the static of internalized truths that tell us that natives wearing hardly any clothes aren't civilized enough to have their own ways of communicating, and listen. To listen beyond the distortions of culture, ethnicity, and class, beyond all the other categories that tell us what something means. And just listen."

"How would the course of history have been changed if Columbus and all who followed in his path had been curious enough or humble enough to listen?" she asked. "How would the future be different if we thought of listening as a political act?"

◊

I thought about listening as an active endeavor, a way to engage in the world that acknowledges the limits of what we know, the need to pay attention, a kind of watchfulness and openness that allows us to hear what was and was not said, to ask questions in a receptive way. I thought again of Minaba and Sepa following their wooden bowl and walking stick across the landscape, listening with their hearts as much as seeing with their eyes, on a pilgrimage to find their way home.

◊

I had come on this trip because on some level I really did believe that gathering stories about land and community could help heal some of the cultural and ecological destruction I saw in the world. In the same way, even with my doubts about the Makalu-Barun Conservation Project, I had believed in Khagendra and the others who worked in the villages to implement the programs. I knew enough about how international conservation and development projects were connected to political and economic institutions that limited what could and could not happen as well as the history of colonialism and domination of developing countries by developed nations that shaped the terms of these encounters. Yet I still believed in Khagendra's ability, as an individual, to make a difference. In some ways—even if not in all ways—I thought that he would be able to transcend or transform the institutions in which he worked and that his actions and, in turn, those institutions could help minimize the exploitation, even if they could not eradicate it. That was how change happened.

But now I was beginning to doubt that such transformation was possible, wondering whether I was simply being naive. Even if Khagendra had interacted differently with the villagers on an individual level, even if the maps had been based on indigenous ways of mapping, he was never going to transform connections based on control into relationships based on openness and reciprocity. The political economy of international development ultimately set the terms of Khagendra's interactions with the villagers. Stories, inspiring as they were, told within that framework were unlikely to bring about the changes that I hoped for. I thought of Cronon's comment about the fences built by the colonists to pen pigs, that it was unlikely that anything I discovered would change anyone's minds, and that the most I could hope was that my research in Hedangna might make people pause, help them see something they had not seen, before doing what they already planned to do.

Similarly, the stories that Brian and I were gathering were meant to encourage people to buy land, to stop felling trees, to donate money, to live more simply. They were not necessarily about telling the whole truth. Sometimes the truth, with its nuances and ambiguities, provided more information than needed, distracted from the objective of the stories—to save more land—and gave people an excuse not to act. At Red Lake, we would have to gloss over the larger story behind the details, not because we were dishonest or our intentions were bad, but because of the objectives of our trip. Like those who worked within the political economy of development, we were working within a cultural economy in which certain ways of knowing—definitive, unambiguous, based on what could be seen and known with the mind—were valued more than others. Although we were at Red Lake to listen, we were ultimately here to listen for only certain kinds of information, the kind of information that reinforced certain ways of knowing.

But just as there are different kinds of power—the power of a healer, the power of a headman or politician—there are different kinds of stories told for different sorts of reasons. There are the stories of *kipat*, or any stories told to accomplish certain goals that defend a particular point of view, whose facts and meaning can change depending on the goal. These were the stories we were on the trip to gather, the kinds of stories I had spent much of my time in Hedangna trying to understand.

Yet more and more, I found that I was drawn to the stories that came to me from the clearing to which the shamans journey, messages received while sitting with plants, images in dreams, the language rising from the darkness. I thought of sitting in the garden at night, surrounded by chamomile and poppies. These images, incomplete and confusing as they were, helped me remember and reinhabit the self who had taken me to that clearing and helped me begin to find my way back on my own. They helped me see, like the shaman, with double eyes, the visible and the invisible—stories, like the *tsawa*, that were doorways into the world of spirit and the sacred; stories that revealed the ambiguities and contradictions of what it means to be alive; stories that touched an essence that endured. These stories were not what Brian and I had come to Red Lake to discover. Yet I was beginning to see that one type was not better or more important than the other. We needed both kinds of stories—those that inspired us to change how we acted in the world and those that deepened our capacity to be present in that world—to heal the destruction we were inflicting on ourselves and the land. No transformation on the land would endure until and unless we transformed as well.

24

THE HEALING STONE

October 1999
COLORADO SPRINGS

After visiting Red Lake, Brian, Avery, Kate, and I headed west across the vast landscape of the Midwest, past fields stretching for miles and huge irrigation machines spraying rows of crops with water. After eating at a diner in a dying downtown in western Kansas and driving through a snow squall as we entered Colorado, we were greeted by a spectacular sunset over the Rocky Mountains.

We were heading to Colorado Springs, notorious for its unregulated development. A small group of dedicated individuals—lawyers and architects, suburban housewives and businessmen—had managed to turn around popular opinion in one of the most conservative cities in the country and raise enough money to protect Stratton Land, one of the last pieces of open space on the front range. The project was significant not just because a grass-roots organization, Cheyenne Commons, had succeeded in protecting this land, but because for the first time in recent history citizens in the city had stood up to oppose development that ignored quality of life. This seemed to be an ideal project for exploring how protecting the land can lead to greater civic engagement. The people with whom I had spoken before we arrived were eager to have their story told, which was a welcome relief after the first two places we had visited.

◊

The morning after we arrived, Avery and I visited Stratton Land. I parked the van by a chain-link fence enclosing the grounds of the high school, at the base of the foothills of the Rockies. Members of Cheyenne Commons had hoped to protect the entire site, which had once been owned by an orphanage, but before they could raise the funds and support, two separate parcels had been sold to developers: one for what was later voted the city's ugliest housing development, and the other for a gated community of 10,000-square-foot homes. Even with all this development so close, Avery and I had not walked very far before the land took on a different feel from the rest of the city. It was quiet, but it was not only that. The landscape—dry, hot, open—was completely different from Sage Mountain in northern New England. But it had a similar quality, a comforting presence, as though we were not alone.

We walked up a buff-colored dirt path, climbed over rocks, and played in an almost dry streambed before climbing back into the deep blue, late September sky. We walked slowly, the pace of an eighteen-month-old, and I found myself imagining that I was walking barefoot across the earth a hundred years ago, the land completely wild, or riding horseback, a cowboy searching for lost cattle, building a cooking fire, and spreading a blanket under the open sky. I imagined moving in ways that did not transform the land as drastically, dramatically, and permanently as it had been changed everywhere, it seemed, except where we were.

The next day, I walked the land with Roberta, a soft-spoken and dreamy older woman dressed in a worn blue T-shirt, blue jeans, and a baseball cap, and Kent, an older man who wore a button-down shirt and spoke in an efficient, clipped tone. Roberta and Kent had helped found Cheyenne Commons. Roberta led us to an entirely different part of the land, which I could not have found on my own because there were no trails or signs. She wanted to see what the local conservation corps had done during a recent work project to make the area more accessible to the public. This was the first time she had seen the city's work, and as we walked Roberta would suddenly interrupt whomever was speaking, even herself, and say with a catch in her voice, "Oh, they're putting a trail here; I thought that this part would be left alone."

I thought of the English-language signs put up by staff of the Makalu-Barun Conservation Project on the trekking route to Makalu, a route that was also traveled by villagers taking their yaks and sheep to the grazing pastures along the Barun River. I often wondered how I would feel to be hiking in the New Hampshire mountains and find signs in a language not my own. I remarked on the distance created by trails and signs, even as they brought more people to the land. They allowed visitors to pay less attention to their surroundings and just follow the trail. Roberta nodded and said that that was her concern as well.

We kept walking and talking. More than anything concrete they said, I was impressed with the way Roberta and Kent interacted. They seemed so different: Kent was matter-of-fact, business-like, and assertive, focusing primarily on the practical aspects of protecting the land, while Roberta seemed whimsical and observant. She talked about walking the land with her daughter and naming different places so they could share what they had seen in those places, even when they were not together. She described the way the land had a spirit of its own. I thought that Kent would dismiss many of Roberta's observations. Instead, he simply nodded, as if she were describing what he felt, but could not express. She treated him with the same respect.

They had come to know each other through the campaign to protect the land, a grueling effort that took more than five years and was carried out by the handful of individuals whom I had met the previous day for lunch at Roberta's restaurant in downtown Colorado Springs. It was a loud, boisterous lunch as the members of Cheyenne Commons tried to tell me their versions of what had happened over the past six years. One would interrupt another, who waited for a moment before jumping back in. One would listen to a few words another said and use them to take the conversation down a different path. Like the members of a family who get along, no one seemed to mind or even notice the interruptions and digressions. Then, too, I had been struck by a quality of respect among them. I asked about it later, and each answered that it was because of Stratton Land—something on that hillside got hold of them, drew them together, and sustained them when the prospects for conservation seemed bleak.

◊

Shortly after we arrived in Colorado Springs, Brian returned to New Hampshire for work. The night before he left, we had a quiet, close dinner alone in a sushi restaurant. He assured me that he was staying true to his promise, that he was fully committed to our marriage. I felt a shift, an opening between us.

He joined us again five days later on a Sunday afternoon. Things had been going well. The project was inspiring. Hearing the members of Cheyenne Commons describe what they had gone through, I felt hopeful. I had made good contacts and lined up important interviews with developers and city planners. I had finally engaged in our project and looked forward to his return.

Brian walked into the two-room tourist cabin where Kate, Avery, and I had been staying for the past week. The cabin was dark and cold and felt cheap. We had tried to make the best of the space, more by joking about it than anything else and by spending as much time as possible outside. The room where Avery and I slept was not neat. My notebooks were stacked on top of my laptop, which was balanced on the tiny circular table, the only one in the room. Clothes were piled on Avery's bed because the dresser drawers were full. The room could have been more orderly. I knew this. When Brian entered the room, he was immediately tense. The tension erupted. We argued. It was not a new argument.

I had placed the smooth green glass, the HEALING stone, from Julie on the windowsill, as I had done in each place we stayed. As Brian angrily unpacked his clothes, he bumped the wall. The glass fell from the sill and broke.

What does it take to belong? Belonging: to be longing. "Longing begets belonging," writes Mary Oak (2000). But the longing to belong kills it.

I kept both parts of the stone.

◊

Although I have pages of notes from interviews—interesting interviews about important issues—most of the time on our trip around the United States was not spent talking to people. It was spent in our large, comfortable green Volkswagen van, driving across the country on this American pilgrimage of sorts. Or it was spent on our own: typing notes, taking photographs, playing with Avery, shopping for food, going for runs, cooking meals, cleaning and packing, getting ready to move to the next place, getting settled in that place, repeating the cycle again.

We were talking about the health of communities and thus about a kind of wholeness. And yet I do not think that we understood, at the time, what it really meant to be whole—what it took to be whole. We never stopped traveling long enough to consider that even though we were talking about wholeness, we were becoming more and more fragmented, in ourselves and with each other, as the trip went on. Nora often spoke about letting the movement of my body start from deep inside, from a cellular level and from my whole self—muscles and bones, head and heart—rather than from just one part of that self from my muscles, say, or, more naturally to me, from my head.

That was not how we moved on this trip, even though we were moving all the time. We were moving because of ideas: of pilgrimage, stories, and home; that the stories mattered; that the trip would save our marriage. Yet the main feeling I had on the trip was numbness, the feeling that comes when we cut ourselves off from movement, not when we hand ourselves over to it.

When Brian, Kate, and I talked, we probably discussed the project—our impressions and plans—and the funny things that Avery had done. Kate does not remember us laughing. I do not remember us talking. What I mostly remember is our silence. Brian and I did not have a lot to say to each other. Although we were together more than we had ever been, the distance between us felt greater than ever before. Every time I tried to talk about the gap, to understand its causes, I simply made things worse. Even if we had been willing or able to be more open with each other, Kate was usually close by. We did not have the privacy needed to get to the bottom of what was wrong. More important, perhaps, neither of us did what was needed to create that privacy. I let myself be absorbed by Avery's needs. Brian let himself be absorbed by the work.

◊

"Soul and habitat are correlates of one another," writes Robert Pogue Harrison (1992:149). The American habitat: four-lane highways; cars speeding across open prairie; mall after mall surrounded by huge, paved parking lots filled with shiny metal cars; tacky tourist cabins built from plywood and furnished with blue acrylic carpets, polyester sheets, and polyester blankets with no weight and no warmth. We shopped in stores where the lights were too bright, especially food stores on Indian reservations, aisle after aisle of sugary cereal and white bread, dozens of each brand; peppers in plastic; apples in wax.

I couldn't touch, couldn't smell, and couldn't taste the food that I was going to put inside me, that I was going to put inside my child, who looked at me with love. Who looked at me with trust. Who trusted me to fill her, in turn, with love. I could not find the essence in the food that makes it last.

In Nepal, as an anthropologist I could look at the physical and cultural landscape from a distance. I was empathetic, of course, and I wanted to do what I could to help the men and women I had come to know. But the conditions in Nepal, as much as they affected me, did not affect me the way conditions in my own country did. I looked at the empty downtowns we passed through, the 10,000-square-foot homes being built in almost every community we visited, and the endless pavement and cars zooming by—juxtaposed with the breathtaking beauty of mountains and rivers, meadows and forests—and I was filled with a hopelessness and despair deeper than I had ever felt. I did not worry that there was another side or perspective that I should try to understand. I did not care about any other side. Something too important, something more important than almost any other thing, was being destroyed beneath all the pavement. That thing—call it life, the essence in life that makes us last—mattered more to me than holding onto the perspective of an anthropologist. Brian felt the same way, which was why he did the work he did. But now, on this trip where our intention was to document how people had changed their lives to protect this essence in the land, we could not find that same essence in each other.

On the pilgrimage to Khembalung, my feet bled. I was hungry and cold, and my load was heavy. In spite of or because of those challenges, I found emotional comfort, closeness, and a kind of openness I had never encountered. I suffered no physical pain or discomfort on our trip around the United States, and yet I experienced more fear, loneliness, anger, and sadness than I had ever encountered.

So much of my cultural understanding of home seems to be shaped by the concept of nostalgia and thus is tinged with a sense of loss and homesickness, the loss of what once was. And our literature is very articulate at registering the effect of that loss on our soul. In contrast, it seems significant that, unlike the Yamphu and so many other cultures, we do not have concepts like *tsawa* and *charawa,* which acknowledge the sacred essence in land and the world around us. This does not mean that we do not have the experience of that sacredness; on our trip, we repeatedly met people whose lives were shaped by that experience. So many of us are groping toward something that we know is real, and yet our response to that longing is often private, drawing us into ourselves rather than connecting us with a larger community in which the experience of inner belonging could resonate with and deepen our sense of belonging in the world.

Stratton Woods was an exception, but primarily because—unlike the Ojibwe at Red Lake and the Yamphu in Hedangna—the members of Cheyenne Commons were lawyers and architects, leaders of the community or members of the same social and economic class as those leaders. The goal of the group was to save a particular piece of land, and the personal connection that individuals in the group felt to the land became something shared, deepening their connections as a community.

Conservation and development projects in Nepal and other developing countries were never simply about conservation. They were ultimately about power: who had rights to resources, what authority gave them those rights, and how the rights were administered. Technical assistance was provided in what usually was a context of unequal rights and power, regardless of whether that context was acknowledged by project administrators.

Conservation projects in the United States inevitably touched on these questions, particularly in communities where certain groups felt they had rights to the resources in question but were marginalized or even excluded from the decision-making process. Yet, unlike the Makalu-Barun Conservation Project, the vision to protect Stratton Land had come from members of the community who then sought technical assistance from conservation organizations. This technical support was, for the most part, simply technical, information passed between members of the same social, cultural, and economic class. It did not involve introducing a view of nature to those with a very different set of beliefs, and the transfer did not occur within a legacy of exploitation.

◊

October 1999
JACKSON, WYOMING

Another run. Sat next to sage, cool wind blowing across open land. Silence in sound. Stillness in wind. The invisible something that sustains us—if and when we let it.

Lying in bed the night before I headed east for the last weekend of the herb apprentice course, I told Brian how alone I felt, how unhappy I was. I thought of Rosemary's observation earlier that summer that depression comes when we are not holding sacred the dream inside. He replied that I had to do more for myself, that I was too pulled between Avery and my work, that I was doing a great job with Avery, but not with myself. As he spoke, I knew that he was right, yet knowing that made me feel even more trapped. The only way out was one more thing to do: take care of myself.

I thought of the night before my conversation with Brian, over dinner with Terry, a writer whose work I greatly respected. She asked about our trip and about the other things I was doing. Mostly, perhaps, she looked at the way I clenched my jaw and the way my eyes did not hold hers, and said, simply, "You are carrying a lot. How do you do it?"

I felt the relief of being seen, a kind of bearing witness, with nothing to do. Yes. I was carrying a lot. I wanted to be held the way I held Avery—to hold myself that way, holding all of myself, not just the parts that were good. And then I could do the things I was called on to do. Brian was also trying to help, and he cared about my happiness in ways that Terry did not. Yet the way Terry expressed her concern was the water and sunlight I needed to again begin to grow.

◊

When I had started the apprenticeship at Sage Mountain, I regarded it as an activity on the side, such as taking a gardening or first-aid course or, in this case, a combination of the two. But the deeper I immersed myself in the program, the more I realized that in many respects I was interested in herbal medicine because it allowed me to continue exploring a way of being in the world that I had experienced in Hedangna. In the philosophy of herbal medicine, I found a resonance with my own views of the sacred, views I had come to understand more clearly in Nepal. Like the ways of knowing at the heart of the Yamphu healers' relationship with the ancestors, herbalism is based on a sense of the sacredness of the earth, a quality of respect and restraint in interactions with the environment, a focus on relationship rather than ownership, and an understanding of the spiritual and cultural dimensions of healing. The practice of herbalism—sitting with plants, harvesting them, and making medicine—allowed me to cultivate the qualities of attention and intention, the two qualities needed to learn to see the visible and the invisible, which were so important to the ways the shamans and priests moved through the world. I wanted to learn to be in the presence of a world we did not make, not just because I wanted to hone my ability to see but, increasingly, because I felt our survival on this earth depended on it.

I embraced these beliefs wholeheartedly because, at that time in my life, more than anything I needed something secure to believe in, a story to carry me along while those I had built my life around fell apart. Ironically, my very immersion in the beliefs and practices of herbal medicine—what I thought was an escape from the distance of anthropology— helped me regain my critical stance. As with the stories of Helen and Scott Nearing, I found myself feeling skeptical and discovered that, in fact, I was most comfortable with one foot on the edge, on the outside, looking in.

That realization, in turn, helped me see that what I had discovered in Hedangna was not a set of beliefs about nature or culture or the sacred. I was not going to find what I was looking for by stepping into another belief system, finding a new cultural truth to replace the old. The act of growing and harvesting plants, making medicines, and learning how to care for my family in such simple ways nurtured parts of myself that had been nurtured in Hedangna. Herbal medicine offered a practice—a practice of wholeness—that embodied those values, and it offered it in a tradition rooted in my own physical and cultural landscape. Most important, though, harvesting medicinal plants cultivated a habit of attention, a way of being in the world that was perhaps more accessible through gathering St. John's wort flowers than through picking up toys strewn across the floor. But what mattered was the attention I brought to each task: how I did whatever I chose to do, not what I chose.

25

THE BLACK BAG

October 1999

SAGE MOUNTAIN

I arrived from Wyoming in the dark, late on a Friday night. It was rainy and cold; the leaves had fallen to the ground; and I felt a part of myself relax, as if I were finally exhaling all the air I had been carrying inside. In part, it was simply the contrast of being away from the tension in the van and the effort of being part of a project and a marriage that I was beginning to realize, especially when I was at Sage Mountain, I no longer believed in. But it was also the way I felt on this mountain, connected to something bigger than myself, where I felt safe opening and speaking from my heart, even if only in the solitude of the woods, but slowly also in the darkness of the sweat lodge, in the care created by Rosemary's presence. I was coming home, which I was beginning to understand was about a fullness within myself, an opening made possible by but not dependent on Sage Mountain.

Herbalist Stephen Buhner had come for the weekend to offer a workshop on sacred plant medicine. He began with a story about fishing with his great-grandfather and about how as a child lying side by side with his great-grandfather on the bank of a river, waiting for the fish to bite, he would feel something pass from his great-grandfather and into himself. That feeling became the thing he wanted more than anything else in the world. "You know that feeling when it is there," he told us. "The energy shifts and you start to pay attention." He paused. "Follow it, and you'll be okay."

He had us close our eyes and led us to find the part of ourselves that was still in touch with the wonder of the world, the part that could experience and understand the spirit of the plants. And then he sent us out to sit in three places: at the side of Rosemary's house, under a maple tree, and next to a boulder at the edge of the woods. We were to list everything we noticed about the rock or the tree or the house and then close our eyes, find the child inside, and ask our inner child to tell us what he or she saw and felt. In that way, he explained, we could begin to learn to intuit the spirit of each place and determine the medicine it had to offer.

I went first to the boulder. I sat on the ground and leaned my back against the hard stone, thinking as I did that it did not look very special: it was just a rock. Even so, I wrote down what I noticed: how the boulder looked and how it felt beneath my fingers and against my back. Then I closed my eyes and searched for the little girl. It took me a while to find her; when I finally did, she was scared and having a hard time breathing. I felt a sudden sense of being crushed in my chest, not being able to find anything inside to resist the constriction from the outside, no heartbeat, just the weight pressing down. I felt as if I were being pushed and then falling through space into blackness, the same sensation I had experienced in a recurring dream as a child, of darkness pressing in. In my teens, I had finally learned that I could escape the dream simply by opening my eyes. The feeling was terrifying, even now, thirty years later, and so I quickly opened my eyes. The impression disappeared. I stood and walked to the maple.

Later, back in the circle, a woman shared her experience while sitting next to the same rock. But she described a feeling of peace, a sense of being grounded and nurtured. I had been so sure that what I had experienced was rock and not me that I described my emotion and asked how it could have been so different from that of the other woman. Stephen explained that I had encountered an aspect of myself that was brought to consciousness by the qualities of the rock, not the rock itself. "We all pick up on the same things," he said, "but we work with them in different ways. You felt the emptiness inside the rock and felt afraid. You will keep getting this lesson," he said, "until you stop and pay attention. Instead of running from it, instead of getting up and walking away, stop," he said. "Say, 'Oh, that's interesting, I wonder what that is about.'"

I did not mention all the emotions that I was running from at the time, most of which were hard and uncomfortable, feelings that I wanted to go away, not about which I was curious, and instead asked, "How can we tell the difference between what is rock and what is our response to rock? What is our projection and what is a way of seeing into the world?" I wondered about this both as a means to interpret my own experience and as an anthropologist. How can anyone raised in a culture shaped by the Enlightenment, a culture that depends on scientific proof to determine what is and is not true, learn to get valid information from intuition and dreams? More than anything, I wanted to trust this intuitive wisdom. But the process seemed too tied up with ego—with ideas about who we wanted to be and what we wanted to see—to be a trustworthy source of information.

"It helps to have a guide," Stephen acknowledged. "But mostly it is a kind of devotional practice. The more self-reflection you engage in, the clearer you will become about what is and isn't you." He then talked about Robert Bly's (1988) "black bag," his description of Carl Jung's concept of the shadow. We put into this bag the parts of ourselves that we do not want to look at, the parts of ourselves that shame us, our flaws and missteps. By the time we reach the middle of our lives, this bag is so long that it drags behind us on the street and gets caught in elevator doors. At some point in our lives, Stephen said, we must open the bag to see what is in it. We will find things that are painful and are hard to accept, but we will also find our creativity. By opening the bag, we begin our journey toward wholeness.

As I walked through the woods during the lunch break, I thought about what Stephen, echoing Bly and Jung and so many others, had said. Bly's metaphor concerned the psychological, but it spoke to the cultural and economic as well. I thought about the sprawling development we kept passing on our trip: acres of pavement, huge houses in gated communities, row upon row of corn, and grocery stores filled with out-of-season strawberries—one of the most toxic fruits available—and blemish-free apples. I thought about the hidden costs of these "advancements": meadows and streams drained and paved over, pesticides polluting soil and water, and people suffering from cancer and poverty. I thought about the unseen impacts of my own actions, no matter how conscientious I tried to be: gasoline for our van and for the airplanes that flew me back and forth across the country, electricity for my computer, and trees felled for paper for my books—the list went on and on. All these unintended consequences stuffed into a huge cultural black bag that let us forget the true cost of our ways of living and imagine that we could continue living as we were and still bring about the changes we hoped for.

Brian's and my trip was to explore how land conservation offered an alternative, providing a path to a healthier culture and economy. Yet as long as the focus was on how the land had been protected rather than on why it needed protection, I did not see any end to the destruction of the land. On our travels and among ourselves, I did not hear any honest discussion of the ways that the choices and actions of all of us—including those concerned about environmental degradation—contributed to the problems we were trying to fix.

At the Watershed gathering of writers, Barry Lopez said, "Streams are messed up, forests are destroyed by an invisible hand. Unless we do something to stop the destruction, ours are some of those hands." This is easy to say, but much, much harder to acknowledge that my hand—not my hand in the abstract, but the hand that held my daughter and dug the soil in which I planted tomatoes— was part of what had created the poison. How could I accept the knowledge that my actions contributed to the smog hanging over the horizon, the clear-cutting of forests, the flooding in southern Asia, the extinction of the polar bears—that I was responsible for the destruction of the things I loved? How could I hold in my body the knowledge that I could not act without creating more poison in the world, take that admission out of my black bag, look at it, and not go numb?

I thought again of our trip in the van across the country and my relationship with Brian. Not only did neither of us want to accept responsibility for the hurtful things we had done, but it was difficult to remember the times that had been good. I tried to recall times we had laughed, the easy spontaneity that came from seventeen years of sharing the rhythms of life, the security of knowing that another person was there, even if he was not always there in the way I hoped. It was hard to acknowledge my part in bringing about the end of something that was—and it was—sacred.

At the center of traditional herbalism are the plants, not parts of plants or constituents of plants, but whole plants. Rosemary once compared scientific research on single plant constituents to determine their medicinal value to looking at only one part of herself: "If you just take the nasty sides of myself and put them out in the world, I'd be a horrible person. But when that's contained, like my toxicity, when that's contained within a whole, it actually works really well."

Rosemary did not mean that a whole plant always has medicine that is good. She meant that medicines derived from whole plants were more balanced than those made from extracted constituents. Stephen seemed to be saying something similar: just as whole plant medicine was more potent and powerful, our own healing depended on becoming whole, on opening and making room for all the parts of ourselves, the bad as well as the good. By making peace with our entire being, we could see the world as it is, not as we wanted it to be to fill a hole inside that we had not filled ourselves.

I was slowly beginning to understand that the crisis in my marriage was a consequence of the same forces behind the crisis in our culture and environment that Brian and I had set out to explore: a fear of seeing that, as Peter Matthiessen said at Watershed, quoting Aleksandr Solzhenitsyn, "the line through good and evil cuts through each of our hearts." Only months later, when I had no other choice but to pack my bags into the same van in which we had driven around the country—this time leaving my home for good—did I realize that coming home, for me and my culture, did not mean settling in a particular community or on a certain piece of land. It meant opening to the reality of my life, stripped of any story to justify what I had done, and seeing that I was responsible for the world in which I lived, the sweetness and the poison, as much by my action as my inaction.

I thought of the deep sense of relief and rest I had felt at dinner when Terry saw not who I wanted her to see, but who I was in that moment, no need to hide. By offering ourselves that same generosity of spirit, making room for the weeds inside as well as the flowers, we might be able to accept responsibility for our actions because that realization would be met with forgiveness rather than judgment. Settling into ourselves, in turn, might allow us to listen with less need to prove, to see with less need to be seen—because we already heard and saw ourselves. I wondered if this might be the path to the cultural changes needed to heal our relationship with the earth.

◊

STEAMBOAT SPRINGS, COLORADO

A few weeks earlier, we had visited ranchers near Steamboat Springs to talk to them and conservationists about their collaboration to protect ranch land in the Elk Valley of Colorado. Sitting at her kitchen table one afternoon, Brian and I spoke with Lynn, a third-generation rancher who had lived in Colorado her entire life, about her relationship to the land and the threats it now faced. She spoke passionately and articulately.

"This is an incredibly beautiful place," she told us. She turned and pointed out the picture window to her husband in the distance, rounding up cattle in the gray-brown pasture of autumn, the pale cottonwoods clustered along the river, the yellow tamaracks on the foothills, and the snow peaks beyond. "One of the reasons it's still really beautiful is because there's been a group of people who have spent their lifeblood working to keep it this way. We love what we do, and this land has us by the throat. . . . You've got to care about it more than anything else in your life, which is what these guys do." She turned back to us and continued:

> You can't legislate morality. You can't legislate love. You can't legislate a sense of caring. If it's not there, it's not there. . . . If we don't imbue somebody else with that spirit of why it's important to keep the land whole, then we're all lost. Because that land doesn't have a chance of going on. . . .
>
> And I suppose it's like loving another person. Somebody that you have the right and the willingness to love and they love you back, then you're going to invest yourself wholly in that relationship.

"The right and the willingness to love." I had never heard it said like that before. I always thought of rights as permission given by one person to another: the right to gather dead wood in a neighbor's forest, to graze sheep in a particular pasture, to inherit land, to vote, to speak. Lynn seemed to be talking about something else.

She continued, "And the land loves you back; it does. If you're allowed to care for it and do what's right, it loves you back by being productive and healthy and taking care of you. Everything you do, every breath you breathe, has to be about maintaining the viability of that land so it can keep going on, giving back for generations to come."

At the end of our conversation, Lynn paused and then added that she and her husband did not own the land outside the picture window, the land she loved so deeply. Not only did they not own it, but after twenty years of working that land, their lease had expired, and they had been asked to leave. They did not know where they would go, she said. They could not afford to buy land of their own in Elk Valley.

The right and the willingness to love. I thought of the overlap between *kipat* and *tsawa,* how a connection in the world of spirit spilled into the world of law. What would it be like if the way we loved something earned us the right to own and use it? What if our willingness to take the risk of loving created the right? Not the other way around. The right to love the land, to love another person, that was not given by someone else, but came from the way we were: trusting nothing but our own presence of body, mind, and soul, and opening utterly to the world around us. That was what gave Ike the right to see Old Ben. That was what opened the door to the world of the ancestors, and maintained the healer's right to be a healer.

What, I wondered, is the cultural cost when we no longer know how to love—love the land, love each other—love as a parent loves, without the need to possess, when the act of loving is the gift you receive?

26

VOICES OF THE LAND

November 1999

NEZ PERCE RESERVATION

After the weekend at Sage Mountain, I met Brian, Avery, and Kate in Spokane, and we drove to the Nez Perce reservation. With the help of a national conservation organization, the tribe had reclaimed some of Chief Joseph's wintering land. We had come to tell the story of that reclamation and what it meant for the tribe.

Avery, always hard to get to sleep, had been especially difficult lately. Often, it took more than an hour for her to fall asleep, and she would wake several times, needing to be comforted and wanting to nurse. It took so long in part because I had been nursing her to sleep, often too tired to try something else, and in part because we were in a different place every week. After an especially difficult night, Brian and I quarreled about what we should do. Avery was nineteen months old, and he thought that she had nursed long enough. In my mind, nursing was connected to the idea of the *tsawa* and the Yamphu drinking from the spring of their ancestors. I felt that Avery should choose on her own when she was ready to leave that spring.

I could not articulate my feelings about nursing and sleep and Avery to Brian in a way that led to any concrete solutions, and I did not have any other suggestions to offer. I simply disagreed with what he said and headed into the cold, windy morning for a run. It was too windy to run, and I was too tired— tired of the movement; tired of the distance between Brian and me, which never grew smaller; tired as the mother of a toddler is always tired. I walked instead.

Too tired even to walk, I lay on the ground by some mullein—with its velvety, pale-green leaves that soothe lungs and tiny yellow flowers that cure earaches—a plant that grows between railroad tracks, along dry dusty banks, in places with no grass and no shade and thus nothing to protect the roots from drought and the leaves from the glare of the sun. Lying there, I had the sense that, like the mullein, I had to grow tall and stately despite the overgrown or undergrown place I found myself in, not letting the place affect how I grew, not letting another's ideas affect my own knowing inside.

◊

A few mornings later, on our last day on the reservation, we sat in Jaime's office. Jaime was one of the younger members of the tribal council. He was a mix of traditional Nez Perce and mainstream American culture: his hair was cut short in the front, with a ponytail hanging down his back; he wore an Indian vest over a University of Oregon sweatshirt. We were staying in his home and one night had taken a sweat in the lodge he had built. The lodge was covered with blankets, as was traditional, but beneath the blankets was insulation. Like the Nez Perce, Jaime was trying to build on what was valuable from the past while creating something new for the future.

That last morning, Jaime said that the land had a voice of its own, but it was being drowned out. Some of us still heard those voices, he continued, but we did not believe what we heard. His words echoed those of Karla, a soft-spoken woman and former member of the tribal council, who had told us the day before that the voices of the ancestors sang when one walked down a valley where they, too, once walked. Those who knew to listen could hear their song.

I knew, again, that I should be skeptical, that I risked falling for stereotypical images of a Native American way, accepting uncritically a romanticized depiction of nature. And yet, traveling across America I had seen so much evidence of a landscape built on the assumption that the earth was not alive, that it did not have a voice of its own, let alone a song. That landscape left me with nothing but despair, alone in our van with a failing marriage. I wanted something else. I wanted to know what would happen if we lived in a way that listened to the voices in the land—of the pine, the oak, the ancestors singing as we walked down the valley where they, too, once walked. What kind of world could we create if we believed that the world around us was alive? I wanted to go further, to ask what to do when the songs my ancestors had sung—not the ancestors I long to have had, but the ones I did have—and the things they had done to the land and the people living on that land were not what I wanted to remember? And I wanted to not be afraid of the answer: "The line through good and evil cuts through each of our hearts."

◊

Driving across the Oregon landscape—Kate visiting friends, Avery asleep in her car seat—Brian and I took off our masks, one by one, deciding that the part of ourselves we had hidden was worth more than staying safe. Each speaking, for once, not of the stories we hoped would be true, but of what was really true. He told me the truth. I told him the truth. No more secrets. Encountering the pain and the anger as we called on a different and deeper part of ourselves, trying to be with what was real.

I opened the door inside for the same reason the shaman makes his offering to the witch, seeing that had I shared what I shared any sooner it would have been for the wrong reasons, because of anger or guilt, not because I realized, finally, that the only thing I had to offer was myself. Not my vision of that self, not Brian's vision, but, simply, that self.

As I opened, finally, hoping that it would be what was needed, deep inside knowing that it was not enough and, even if it had been, that it came too late.

To accept the past. To see, sometimes, that the past holds too much hurt to restore what we long to restore, that we cannot return to the clearing in the woods. To start where we were, in a broken place—I thought of the woods at Sage Mountain after the ice storm, the twigs broken from the trees, branches strewn across the forest floor. But to imagine anything else was to imagine what was not true.

I went for a run on a needle-covered path along a river and stopped by a huge sequoia. I wrapped my arms around the trunk and asked how to be in this place where the road I had seen stretching before me had suddenly disappeared and the story Brian and I had told each other and ourselves had come to an end. What felt like an electric current flowed from the sky through the tree, touching the raw, vulnerable self I had never touched, this first time of speaking that self to the world.

Climbing to the caves barefoot.

◊

When we had gone to look at the wolves on the Nez Perce reservation, Black Beaver had described a morning when he had been preoccupied with his own finances and worries from home while he was inside the fence working with the wolves. Not paying attention, he did something that angered one of the wolves; she bared her teeth, bristled her fur, and looked ready to attack. The wolf handler told Black Beaver, "We can handle this in two ways: either leave or ride it out." Black Beaver said he wanted to ride it out. So the wolf handler told him to bend his head down and act apologetic, giving into the power of the wolf. Black Beaver did as he was told. Once the wolf's authority was recognized, she backed off. He learned his lesson, Black Beaver said. When he could not be fully present, he did not enter the territory of the wolves.

How do we act in the presence of what we cannot fully understand or control? What does it take, in other words, to be in the presence of wildness? Where do we place our trust?

When faced with forces that for them are no less fearful—demons within and demons without, terrors almost too great to bear—the shamans and priests in Hedangna hold their ground. They wave their bamboo wands. They chant "*Namaste!*" and "*Sha'de!*" (I greet the God within you!), and they chant it again and again and again. They try to control what they can, and they surrender utterly to what they cannot.

27

THE WATERFALL

December 1999

I was in the Volkswagen van driving through Oklahoma. It was dark. The moon, a full December moon, huge and yellow, rose through the darkening sky. We were heading east after four months on the road. Brian and I had just reunited after three weeks apart. Kate had gone home. Avery slept in the back. Brian was at the wheel. I could not stop staring at the moon. It was so big. It gave me hope, so I talked about Caroline. I told Brian about the very first day I had seen her, how I lay on her table, and, after silently placing her hands on my body for a long, long time, she had said so quietly that I could hardly hear, that she could not find me.

"Where are you?" she asked.

"I don't know," I whispered. "I don't know."

We continued to drive, watching the moon. It felt risky to share this with Brian. Whenever I tried before, I had not been able to express myself in a way that he could understand and that made him want to understand. Plus, there was now this distance between us. But he seemed to be listening, so I kept talk-ing. I described another day when Caroline had placed her hands on my abdo-men, and the image of sea urchins, sea cucumbers, starfish, and an anemone had suddenly appeared, crowded into my insides. A tidal pool right there in the space I thought was empty. "Do you know how much is inside?" she asked. "Do you see what is there?" The tidal pool vanished before I had time to respond.

I glanced at Brian at the wheel, trying to sense his interest. On the dash-board, I placed the turtle from my sister, the green glass from Julie, and a wolf I had bought for Brian. I left them in the moonlight and talked about the bridge I had seen as Caroline placed her hands on my clavicle and shoulder blade, a bamboo bridge high above a ravine in Nepal, a bridge stretching through my back. I was crossing the bridge, placing my feet carefully, one after the other on the single bamboo pole, hands gripping the sloping railings on either side. And then the bridge snapped, and I was falling into the river below. The cur-rent was strong. I began to sink; I was drowning and then struggling on the surface to get some air. And then the image disappeared.

I paused. I took out a shell I had picked up on the beach at Lake Ozette and burned some sage. A burning ember fell on the seat. Brian turned and looked sharply at the hole that it made. But I was determined to let in the moonlight. I was determined to not disappear. I kept talking.

It was 1:00 A.M., the same night. We had stopped at a motel somewhere between Oklahoma City and Little Rock. I put Avery to sleep in one bed. We climbed into the other. It was odd, sleeping in the same bed. I was not sure what to do, not sure where we were. I thought about an intimacy in my body that came without question. I thought of the gap between what my body was used to and what my mind now knew. I was tired, so I just went to sleep. I awoke from a dream to find Brian touching me with a gentleness that was new, an intimacy I had longed for. It was such a surprise—like an unexpected dream. This sense of being seen. It made me want to see. Then something shifted. I realized that he, too, was waking from a dream, that he, too, was sur-prised. I felt him disappear even as he held me in his arms.

I wondered which is worse, not being seen, or being seen as someone else.

◊

Thanksgiving 1999
MOLOKAI, HAWAII

We were on Molokai, looking at an effort to bring back the traditional irrigation systems in the Halawa Valley, the first valley settled in Hawaii. This turned out to be the last stop of the trip. Glenn, a tall, bearded Hawaiian, led us through the jungle. He said that he was taking us to see the irrigation channels that had been brought back into use and the newly planted taro, on land that was being farmed as his ancestors had worked it, but we passed the channels and taro long before. It was his first time back in the valley after a stroke that had erased his memory and, for the time being at least, brought to a halt the work that had brought us to the valley. We had come anyway and, as usual, were looking for a story.

He walked up the path through land that was once his home as though he had never been here or had been here long ago. We reached an opening in the trees and saw a waterfall at the end of the valley, plummeting down the rock face and disappearing into the jungle. Glenn looked at it, clearly disoriented. He turned to us, suddenly, seemingly not sure who we were, forgetting why we had come. He was silent for a few moments and then asked in a quiet voice, "How did we get here?" We were silent as well. "You led us here," Brian said, trying to help him through his confusion. "We have come to learn about your work, about what you have been doing to bring this land back alive, how you have been farming it as it has always been farmed." Glenn had asked the one question I had been asking myself, again and again, for the past four months. His confusion shook me deeply.

Later that day, Brian, Avery, and I walked back up the valley, trying to make our way to the waterfall. It was too far for Avery to walk, and so we decided to take turns heading up the valley on our own. I went first, but Brian asked me to come back quickly so he could get to the waterfall in time to take pictures while it was still light.

I walked silently along the dirt path. The trail divided. I picked one path over the crumbling terraces, farmed years earlier by men and women whose spirits, it was said, continued to haunt the valley. I scrambled over rocks. I moved quickly, not noticing where I was, not noticing where I had been. The trail disappeared. I assumed that I had only to follow the stream, but the rocks were too big. I did not have enough time. The way was not clear, and so I gave up. I bushwhacked through the jungle, back over the overgrown terraces, until

I again found the trail. Brian then headed up the trail, returning several hours later. He had made it to one of the waterfalls. But as he tried to photograph it, his camera fell off the tripod and into the water below.

◊

Three days later, I returned to the valley. This time I was alone. The night before, we had changed our plans. We would end the trip early and send Kate home. Brian would go to New Mexico to do the rest of the research on his own. I would head to Seattle, to visit my sister. When we returned to New Hampshire, in a month, we would live apart. He needed time and space, he said. He had to be on his own.

I was stunned at how events had unfolded; I thought that we were headed in a different direction, that we were going to try to work it out despite all that had happened. I started up the valley, pausing to place an offering of tiny, purple flowers on a rock. I asked that I be big enough and open enough to receive what was there to receive and humble enough to not ask for more. My hand throbbed from where I had punched a wall the night before. I walked as the Yamphu say their ancestors had walked, searching for a place to settle. Walking, walking. Watching, watching. Looking, looking. Paying attention to what was there. Walking, walking. Watching, watching. Not knowing where I was going.

I reached the bend in the stream where I had lost the trail three days earlier. The flap of wings turned my attention downstream, where I noticed a small trail leading from the other side of the stream. I crossed the stream and followed the trail. It climbed around the bend, and I saw a torrent of water pouring into a large pool surrounded by rock. The water flowed out at the base, becoming the stream that once watered the fields down the valley. Cliffs rose from either side of the pool.

I was breathless. I climbed over boulders and stood at the edge of the pool. I took off my clothes. I walked slowly into the water. I kept looking over my shoulder, although I was not sure why. I swam, quickly, into the center. I started to circle, slowly, around the center, and then continued to swim to the base of the waterfall. The wind blew the water into my face; it took my breath. I shut my eyes, swimming quickly. I reached the waterfall and drank several handfuls of the water tumbling over the cliff. I turned and swam quickly back to the other side. I climbed onto the rocks and pulled on my clothes. I felt the tension of the last year drain from my body. In that moment, I knew that I would be okay. Deeply grateful, I bowed my head and then turned and headed back down the trail.

28

BARE FEET

Make a list. A rock altar in the woods, a garden, a glass vase, photographs of Hedangna, long forks, wooden spoons, a wineglass or two or three—enough to have guests—chairs for them to sit on, someone to make the coffee, to wash the dishes, to walk with to the lake, to sit with on the stoop, to call me in the middle of the day. Someone to remember with: the smoky smell of Hedangna, the taste of warm *jad* in Dhanmaya's house.

What brings two lives together? What holds them there? What happens when the sounds, smells, and images that get inside are inside just you, and no one is there to recall them? The path to the past was lost, and I had been told to make a list of items that I wanted to keep, objects of a shared life, a divided life, objects that were put on the porch in the night, and the door to the house was locked.

◊

Being in Hedangna prepared me for this time of sitting cross-legged on the floor of the tiny kitchen in a rented house and eating a bowl of pasta with my two-year-old daughter. Singing thanks to the mother earth, holding hands as we ate, drawing lions on the chalkboard for entertainment, lighting a candle when I could find the matches to guide me through this thin place, the thinnest I had entered.

◊

July 2000

Avery was crying. We were driving to the country store to meet Brian so she could go to his house for three days. These transitions were always hard, for me and for her. She did not want to go. I did not know what to say. I did not want her to go either; more than anything, I did not want her to go. But I could not tell her that, so instead I talked about her heart. I talked about mine. I told her that when she was not with me, I could look inside and find her there, in my heart. I said this even though I did not always trust that it was true. I told her to do the same. I said that her father loved her, too, and that he also wanted to see her. I reminded her how much fun they had when they were together. Tears slipped from her eyes. I sang a song. I told her that he would be waiting.

We reached the store. She climbed out of her car seat, knowing that she had no choice. He picked her up and carried her to a car that once held my presence, now replaced by another. I climbed back into my car and, without looking back, drove home, the memory of her eyes in the rear-view mirror as she learned the lessons of lines.

◊

As I drifted off to sleep, I asked about numbness. I asked about the thick feeling in my heart, about fingers that touched but could not feel, about blood drawing inward because the world seemed too cold. That night, I dreamed about moving, about kitchens and new houses, worrying if the new floor would be easier to clean than the old. I dreamed about boxes. I could do this, I thought, if I were dancing. I would know that numbness is just the shadow of stillness; sorrow, the dark side of flow. My body would move, and in the movement I would let myself feel because I would trust that eventually the music would change. Without the music, it was harder to trust. It was harder to keep moving or harder to stop. Support precedes movement, Nora always said. Except once, when she added that sometimes we just have to fall.

◊

October 2000

I was attending a presentation in the Department of Environmental Studies at Dartmouth College, where I taught the course "Gender and the Environment." I was trying to pay attention to a talk about research on changes in reindeer herding and computer modeling, but did not see the connection. I was taking notes, hoping that by writing I would begin to understand the meaning.

Julie had just died. I was fighting for my rights to spend four days out of seven with my daughter, my two-and-a-half-year-old daughter. I had moved five times in the past ten months.

I continued to take notes, doing what had worked in the past, doing what no longer worked in the present: impacts and systems, community and dialogue. A language I no longer knew. What I knew was what it took to pack my car with what it would hold and leave my home, because the man with whom I had lived for fourteen years, the man who was the father of my child, wanted to try something new. What I knew was what it took to wake in the middle of the night because my child had awoken in tears, because my child needed me to hold her hand and sing a song, and to rise several hours later and walk into the unknown, clutching the hand of a two-year-old girl. What I knew was what it took to shovel dirt onto the casket of a thirty-seven-year-old woman, shovel backward so that God did not think it was easy.

A language I no longer knew. The boundaries between worlds blurred, and suddenly I was a woman, a villager, at a meeting in Hedangna. I heard a man, a Nepali, from Kathmandu. He was dressed in a warm jacket and heavy boots. He looked well fed. He was talking. I heard terms like "user groups," "community development," "local participation," and "income generation." He said that he cared about our ideas. He wanted us to participate. He continued to talk. I did not understand his words. I could not understand his meaning. I looked instead at the yellow nylon tent where he slept and at his black-and-blue knapsack. I looked at the porter who carried and cooked his rice. I wondered what else they would eat with the rice.

He was still talking. I no longer knew what he was talking about. What I knew was the tiredness in my legs, the searing pain from the cuts in my bare feet, the ache in my chest as I coughed—the ache in my chest, always that ache, as I remembered the children, my children, whose bodies I had borne and buried.

There was rice to dry at home. There was rice to dehusk at home and rice to cook. There was water to fetch and dishes to wash. There were children, children who were still alive and needed care. I picked up my basket and left. I picked up my basket to go home. I missed out on the promise of something better. I remained instead with what I knew.

◊

July 2000

I was back in the summer, before Julie died. I called her, as I did each day. She did not want to talk to me, did not want to talk to anyone, especially on the phone. The cancer had spread to her brain, and she was about to start radiation. She was angry, angry with her mother, her doctors, her sisters, and me. I told her that I would drive down to Massachusetts to see her, but she said, "No, Ann, don't," in a tone that meant it. I no longer knew what to say, even if she were willing to talk. We hung up.

Later that evening, I lit candles and made an altar on the wooden floor beneath the window. The room was bare. Bare because I had no furniture, bare because I did not want any. I spoke out loud as I gathered stones and shells, flowers I had picked in the meadow, candles, and scarves and arranged them on the floor. I told Julie how much I loved her and how I wanted to help; I told her things that she did not want to hear, how much I would miss her when she was gone, all the things I had said about what her friendship meant. Alone in the dark, I said things that I had been afraid to say out loud, wondering about her pain, her fear of what was coming, and her sadness. And then I began to dance. I danced and I danced and as I was dancing, I suddenly found Julie. I had not expected this. But I was grateful, deeply grateful, that she had appeared. She was lying in her bed, sleeping. I lifted her as gently as I could and carried her to a clearing in the woods covered in moss. I placed her on the moss. She was surrounded by pink lady slippers, the lady slippers I had seen earlier that day as I was running through the woods. The paper-thin petals already were tinged with brown, and the sac was so full and so delicate that the flowers reminded me of Julie, reminded me of how beautiful everything is and how quickly that beauty slips away.

I knelt next to her and placed my hands on her, as Nora had shown me: nothing to do, no place to go. I did this for what seemed like forever, until suddenly I sensed that it was time to be done. I lifted her just as gently as before and carried her through the night. I placed her carefully in her bed, pulling the covers to her chin. I bowed deeply, turned away, and then traveled home through the darkness.

A memory as real now in my mind in a misty sort of way as when I had laid my hands on her, cellular holding: my offering when there was nothing to say.

◊

September 2000

I moved into my third house, by the third lake. There was one message on the phone when I arrived, from Julie, sending her love, not wanting me to feel alone.

◊

October 2000

Again, I was driving Avery to the country store to meet Brian. Again, she was sad, and I was looking for a distraction. I pointed out the window to the stumps of trees that had just been cut. We had just read *The Lorax*, Dr. Seuss's story about the cutting of the trees, and I said something about how sad it was that the trees were gone. Avery shook her head. "It isn't sad," she replied. And then she said that it was like when I told her that I was in her heart, even when I was not present, even when she was with Daddy. The ability to see what is visible and what is not.

Another day, Avery took the two pieces of the HEALING glass from my altar and pressed them together, showing me what was needed to again make it whole.

I was lying in a meadow on a hill that overlooked a farm in Vermont, watching clouds race across the sky. I asked the land what it wanted. It was autumn. The sun warmed my skin, and the grass was brown and scratchy against my back. It rained. The sun returned. In three months, this ground would be frozen, covered in snow. And in another three months, it would be warm again, carpeted with new grass. The plants grow when the world is inviting; they retreat when it is not—a different sort of music. This was the music I could move to.

◊

January 2001

I went to sleep in the third and last house by a lake, an open notebook by the side of my bed. I dreamed more clearly than I had in months. I awoke in the dark to Avery crying, softly, in her bed next to mine. I lay next to her. Her cries got louder. I brought her into my bed. She refused to lie down. Refused to be held. Refused a song or a story. She continued to cry. I lay down. She remained sitting, pushing my arm away. I longed for sleep, for hers and for mine. I longed for space, space to be myself and to tell whatever story was in me to tell. I tried to remember my dreams, but they were slipping away. She continued to cry. She cried louder when I shut my eyes. So I opened them. My lungs tightened. I did not know how to reach her. I did not know if I had it in me to try.

I took a breath and began to chant. I told her that I loved her. I told her that her father loved her. I realized that I had not mentioned him all week. I repeated, Daddy loves you. And, again, Daddy loves you. I moved on. Whiskey loves you. She wiped her fist across her eyes. Granddaddy loves you. Molly loves you. Sally loves you. Her cries got quieter. Kate loves you. Zoe loves you. She lay her head on my chest. Emma loves you. Henry and Paul and Cornelia love you. She asked for a song. I sang a song, softly: "Puff the Magic Dragon." By the end of the first verse, she had fallen asleep. I lay awake, the weight of my child's head on my chest. I thought about movement and space and light; I felt the weight on my chest. My lungs began to expand. I always left, before—left my home, my marriage, and my culture—thinking that I would find space and air somewhere else and telling myself that openness was a place, not a way of being in place, any place, especially the one where I was. I felt the weight on my chest, the weight, weight of love, inside, pulling me out of my dreams, helping me find my way home.

I had traveled halfway around the world, all the way across my own country, looking for a place in the land to come home to. If only I found that place, I had thought, everything would fall into place. It was only in this year of moving from house to house that I came to realize—really realize—that home is not a place we ever reach. It is those moments—in time, not in space—where the wind cannot reach, the eddies and pools where things do not tremble. It is less a noun than a verb, an attitude and an action: living from the inside out, following your own thread for no other reason than that it is your thread and you have only one time around to follow it.

I finally understood that there was a knowing in my bones and in my body that I had thought was in the bones and bodies of only the men and, especially, the women who lived in a far-off place. I now knew that I, too, could pick up the hoe and return to the fields. Again. And again. Alone, if I had to. With company if I could. In joy and in sorrow: bare feet on wet earth.

BIBLIOGRAPHY

Bly, Robert. *A Little Book on the Human Shadow.* San Francisco: Harper & Row, 1988.

Calasso, Roberto. *The Marriage of Cadmus and Harmony.* Translated by Tim Parks. New York: Vintage Books, 1993.

Cronon, William. *Changes in the Land: Indians, Colonists, and the Ecology of New England.* New York: Hill and Wang, 1983.

——. *Nature's Metropolis: Chicago and the Great West.* New York: Norton, 1991.

Dante Alighieri. *The Divine Comedy.* Translated by John D. Sinclair. New York: Oxford University Press, 1981.

Eliot, T. S. *Four Quartets.* New York: Harcourt Brace Jovanovich, 1971.

"The Encompassing Web." *Ecologist* 22 (1992): 149. [Special issue: "Whose Common Future?"]

Faulkner, William. *Go Down, Moses.* New York: Vintage Books, 1973.

Harrison, Robert Pogue. *Forests: The Shadow of Civilization.* Chicago: University of Chicago Press, 1992.

Leopold, Aldo. *A Sand County Almanac, and Sketches Here and There.* New York: Oxford University Press, 1987

Lopez, Barry. *Crow and Weasel.* New York: Farrar, Straus and Giroux, 1998.

Nearing, Helen, and Scott Nearing. *The Good Life: Helen and Scott Nearing's Sixty Years of Self-Sufficient Living.* New York: Schocken Books, 1989.

Oak, Mary. Essay from Centrum Writing Workshop with Terry Tempest Williams. Port Townsend, Wash., 2000.

Orofino, Giacomella. "The Tibetan Myth of the Hidden Valley with Reference to the Visionary Geography of Nepal." *East and West* 41 (1991): 239–271.

Regmi, Mahesh C. *Land Tenure and Taxation in Nepal.* Kathmandu: Ratna Pustak Bhandar, 1978.

Robinson, Marilynne. *Housekeeping.* New York: Bantam Books, 1980.

Saint-Exupéry, Antoine de. *The Little Prince.* Translated by Katherine Woods. New York: Harcourt Brace Jovanovich, 1971.

Snow, Donald. "Save the Forests, Store the Floods." *Northern Lights* 7 (1991): 12–17.

Thompson, E. P. "The Grid of Inheritance: A Comment." In *Family and Inheritance: Rural Society in Western Europe, 1200–1800,* ed. Jack Goody, Joan Thirsk, and E. P. Thompson, 328–360. Cambridge: Cambridge University Press, 1976.

Whyte, David. *The Poetry of Self-Compassion.* Langley, Wash.: Many Rivers, 1991. [Compact disc]

Woolf, Virginia. *Orlando: A Biography.* Ware, Eng.: Wordsworth Editions, 1995.

DATE DUE

DEMCO 38-296